MAGNETO LUMINOUS
CHEMICAL VAPOR
DEPOSITION

GREEN CHEMISTRY AND CHEMICAL ENGINEERING

Series Editor: Sunggyu Lee

Missouri University of Science and Technology, Rolla, USA

Proton Exchange Membrane Fuel Cells: Contamination and Mitigation Strategies

Hui Li, Shanna Knights, Zheng Shi, John W. Van Zee, and Jiujun Zhang

Proton Exchange Membrane Fuel Cells: Materials Properties and Performance

David P. Wilkinson, Jiujun Zhang, Rob Hui, Jeffrey Fergus, and Xianguo Li

Solid Oxide Fuel Cells: Materials Properties and Performance

Jeffrey Fergus, Rob Hui, Xianguo Li, David P. Wilkinson, and Jiujun Zhang

Efficiency and Sustainability in the Energy and Chemical Industries: Scientific Principles and Case Studies, Second Edition

Krishnan Sankaranarayanan, Jakob de Swaan Arons, and Hedzer van der Kooi

Nuclear Hydrogen Production Handbook

Xing L. Yan and Ryutaro Hino

Magneto Luminous Chemical Vapor Deposition

Hirotsugu Yasuda

MAGNETO LUMINOUS CHEMICAL VAPOR DEPOSITION

HIROTSUGU YASUDA

CRC Press
Taylor & Francis Group
Boca Raton   London   New York

CRC Press is an imprint of the
Taylor & Francis Group, an **informa** business

CRC Press
Taylor & Francis Group
6000 Broken Sound Parkway NW, Suite 300
Boca Raton, FL 33487-2742

First issued in paperback 2017

© 2011 by Taylor and Francis Group, LLC
CRC Press is an imprint of Taylor & Francis Group, an Informa business

No claim to original U.S. Government works

ISBN 13: 978-1-4398-3877-8 (hbk)
ISBN 13: 978-1-138-07209-1 (pbk)

Visit the Taylor & Francis Web site at
http://www.taylorandfrancis.com

and the CRC Press Web site at
http://www.crcpress.com

Contents

Preface...xi
The Author xiii

1. Introduction 1

2. Context of Terms Used and Concepts 5
   2.1   Plasma...5
         2.1.1   Equilibrium and Nonequilibrium Plasma 5
         2.1.2   Low-Pressure (Low-Temperature) Plasma 6
         2.1.3   Deposition Plasma and Nondeposition Plasma.................6
         2.1.4   Reactive Plasma and Nonreactive Plasma 6
         2.1.5   Plasma Sheath...6
         2.1.6   Magneto-Plasma 7
   2.2   Electric Discharge of Gas 7
         2.2.1   Direct Current (DC) Discharge..7
         2.2.2   Audio Frequency (AF) Discharge...7
         2.2.3   Radio Frequency (RF) Discharge..7
         2.2.4   Cathode Fall 7
         2.2.5   Ionization Glow 7
         2.2.6   Dissociation Glow 8
         2.2.7   Self-Bias and Bias Potential 8
         2.2.8   Deposition Gas 8
         2.2.9   Nondeposition Gas 8
         2.2.10 Electronegativity 8
   2.3   Gas Phase 8
         2.3.1   Gas Equation...8
         2.3.2   Flow Rate, Volume Flow Rate, Mass Flow Rate.................9
         2.3.3   Ionized Gas 9
         2.3.4   Luminous Gas 9
         2.3.5   Photo Emitting Excited Neutrals...9
         2.3.6   Magnetic Field in Gas Phase 9
   2.4   Polymerization and Material Formation 10
         2.4.1   Plasma Polymerization 10
         2.4.2   Chemical Vapor Deposition (CVD) 10
         2.4.3   Plasma Chemical Vapor Deposition (PCVD).................... 10
         2.4.4   Plasma Enhanced CVD (PECVD); Plasma Assisted
                  CVD (PACVD) 10
         2.4.5   Luminous Chemical Vapor Deposition (LCVD) 10
         2.4.6   Magnetron Plasma Sputter Deposition 11

        2.4.7   Magneto-Luminous CVD, Magneto-Plasma CVD,
                 Magnetron-Plasma Polymerization 11
        2.4.8   Material Formation Mechanism 11
        2.4.9   Deposition Mechanism 11
   2.5   Surface and Interface 11
        2.5.1   Surface 11
        2.5.2   Interface 11
        2.5.3   Interfacial Electron Transfer 12
        2.5.4   Surface Static Charge 12
        2.5.5   Molecular Configuration 12
        2.5.6   Surface Configuration 12
        2.5.7   Surface State 12
        2.5.8   Imperturbable Surface State 12
   2.6   Biocompatibility 13
        2.6.1   Biocompatibility of Artificial Material 13
        2.6.2   Sustainability of Biocompatibility 13
        2.6.3   Minimum Perturbation Theory of Biocompatibility 13

3. Green Deposition Coating of Nanofilms 15
   3.1   Front-End Approach and Rear-End Approach in Green
       Processing 15
        3.1.1   Low-Pressure Magneto-Luminous Chemical Vapor
                 Deposition (MLCVD) Coating 16
        3.1.2   Layer-By-Layer (LBL) Coating 16
        3.1.3   Cost of Coating Processes 18
   3.2   System Approach Interface Engineering with Green
       Processes 21
        3.2.1   System Approach Interface Engineering for
                 Corrosion Protection 21
        3.2.2   Green System Approach Interface Engineering by
                 Magneto-Luminous CVD 24
   References 24

4. Plasma Phase and Luminous Gas Phase 27
   4.1   Plasma Phase 27
   4.2   Luminous Gas Phase 28
   References 30

5. Dielectric Breakdown of Gas Phase 33
   5.1   Significance of Dielectric Breakdown of Gas on Luminous
       Chemical Vapor Deposition 33
   5.2   Breakdown of Nondeposition Gas and Deposition Gas 35
        5.2.1   Dissociation Glow and Ionization Glow 35
        5.2.2   Interrelationship of Dissociation Glow and
                 Ionization Glow and Influence of Power Source 42

5.3 Source of Electrons for Electron Avalanche to Cause Gas
Phase Breakdown..44
5.3.1 Secondary Electrons Emission by the Bombardment
of Accelerated Ions Caused by the Ionization of Gas
in the Gas Phase; Townsend's Gas Phase Ionization
Hypothesis..44
5.3.2 Free Electrons Emission from the Surface State
of Cathode Metal by the Applied Electric Field;
Primary Electron Emission Principle by Yasuda 45
5.4 Interfacial Electron Transfer 46
5.4.1 Interfacial Electron Transfer in the Static Charge
Creation in the Contact Electrification................................46
5.4.2 Correlation between Number of Free Electrons
in the Surface State of the Cathode Metal and the
Breakdown Current 48
5.5 Experimental Examination of Gas Phase Breakdown...................51
5.5.1 Parameters That Influence Gas Phase Breakdown 51
5.5.2 Breakdown Voltage According to Townsend–
Paschen Hypothesis..52
5.5.2.1 Townsend Theory of Dielectric Gas Phase
Breakdown 52
5.5.2.2 Flaws in Paschen–Townsend Treatment of
Gas Phase Breakdown 54
5.5.2.3 Important Factors Not Considered in
Paschen–Townsend Interpretation of the
Gas Phase Breakdown 55
5.5.2.4 Breakdown Voltage According to the Dark
Gas Phase Parameters..............................55
5.5.3 Breakdown Process Investigated by Breakdown
Voltage, Breakdown Current, and Derived Parameters ...59
5.6 Factors That Control Transformation of Gas Phase 61
5.6.1 Parameters of Reaction Kinetics 61
5.7 Electronegativity of Atom and Efficiency of Electron-Impact
Reactions 62
5.8 Gas Phase Breakdown as Functions of the System Parameters 63
References 70

6. Influence of Magnetic Field on Luminous Gas Phase 71
6.1 Influence of Magnetic Field on Roles of Electrons 71
6.1.1 Changes in Distribution Profile of T_e and N_e in
Luminous Gas Phase...71
6.1.2 Magnetron Cathode (MC) and Nonmagnetron
Anode (NA)..73
6.1.3 Nonmagnetron Cathode (NC) and Magnetron
Anode (MA)..75

6.2   Shaping of Negative Glow Near the Magnetron Anode 76
        6.2.1   Negative Glow of Argon 76
                6.2.1.1   Reexcitation of Photo-Emitting Species by
                            Low-Energy Electrons 77
        6.2.2   Influence of Magnetron Anode on Glow
                Characteristics of Deposition Gas 78
        6.2.3   Shift of Dissociation Glow from Cathode Surface to
                Gas Phase 81
6.3   Influence of Magnetic Field on Dielectric Breakdown of Gas
        Phase 81
6.4   Electrons in Electric Field and in Magnetic Field 88
6.5   Implications of Magnetron Gas Phase Breakdown 90
        6.5.1   Magnetron Chemical Vapor Deposition versus
                Magnetron Sputtering of Cathode Metal 90
        6.5.2   Magnetron Discharge Sputtering 92
6.6   Magnetic Field Initiation of Luminous Gas Phase 94
        6.6.1   Collisions of Gas Molecules with Electrons in
                Magnetic Field 94
        6.6.2   Potential Mechanism for Inception of Aurora Borealis 96
References 98

7. **Polymer Formation Mechanism in Luminous Gas** 101
7.1   Free-Radical Polymerization and Free-Radical Polymer
        Formation in Luminous Gas Phase 101
7.2   Repeating Step Growth Polymerization (RSGP) Mechanism 102
7.3   Competitive Ablation and Polymerization (CAP) Principle 112
7.4   Influence of Unaccounted Factors 116
        7.4.1   Substrate Material 116
        7.4.2   Reactor Wall Contamination 118
7.5   Dissociation of Monomer Molecules 122
7.6   Dependence of Polymer Formation on Operation Parameters ... 124
References 127

8. **Operation Parameters and Deposition Kinetics** 129
8.1   Operation Parameters and Plasma Polymerization Process 129
        8.1.1   Operational Parameter That Influences Repeating
                Step Growth Polymerization (RSGP) Mechanism 129
        8.1.2   Flow Rate and System Pressure of Gas in General 134
        8.1.3   Control of Monomer Flow Rate and the System
                Pressure of Plasma Polymerization Reactor 135
8.2   Deposition Kinetics 136
        8.2.1   Mass Balance in Flow Deposition System 136
        8.2.2   Normalized Energy Input Parameter to Luminous
                Gas Phase 138
        8.2.3   Normalized Deposition Rate 142

8.3   Properties of Plasma Polymers and Domains of Plasma
        Polymerization.. 144
        8.3.1   Type A and Type B Plasma Polymers 144
        8.3.2   Utilities of Type A and Type B Plasma Polymers............ 153
8.4   Partition of Deposition on Electrode and Deposition on
        Surface in Gas Phase.. 154
        8.4.1   Cathodic Polymerization versus Polymerization in
                Negative Glow of Direct Current (DC) Discharge.......... 154
        8.4.2   Deposition Profile on Electrode........................... 154
        8.4.3   Deposition on Floating Substrate in Luminous Gas
                Phase... 159
        8.4.4   Role of Anode in DC Cathodic Polymerization 162
        8.4.5   DC Plasma Polymerization in a Closed System.............. 165
References 168

9. **Magneto-Luminous Chemical Vapor Deposition**................. 171
   9.1   Domain of Magneto-Luminous Chemical Vapor Deposition
         (MLCVD) 171
   9.2   Toroidal Glow Surface without Deposition.................... 175
   9.3   Confined Luminous Gas Phase in Low Pressure........... 177
   9.4   Polymer Formation and Deposition in Low Pressure 178
         9.4.1   Deposition Kinetics in Domain M..................... 178
         9.4.2   Pressure Dependence of Deposition Rate 182
         9.4.3   Small Grain Size and Uniform Smooth Surface............. 184
         9.4.4   Strong Adhesion to Substrate Surface................ 185

10. **Applications of Magneto-Luminous Chemical Vapor Deposition** ... 187
    10.1   Implantation of Imperturbable Surface State on Substrate......... 187
           10.1.1   Surface Dynamic Change.............................. 187
           10.1.2   Factors Involved in an Interface...................... 189
           10.1.3   Molecular Configuration versus Surface
                    Configuration 190
           10.1.4   Implantation of Surface State of Magneto-Luminous
                    Chemical Vapor Deposition (MLCVD) Nanofilm
                    onto Material Surface 196
    10.2   MLCVD Nanofilm for Biocompatibility 200
           10.2.1   Imperturbable Surface State and Biocompatibility......... 200
           10.2.2   Encapsulation of Silicone Contact Lens.............. 201
                    10.2.2.1   Dye Penetration Test 203
                    10.2.2.2   Effect of Coating Thickness 205
                    10.2.2.3   Effect of Power Input Level................ 207
                    10.2.2.4   Overall Effects of MLCVD Coating........... 209
           10.2.3   Encapsulation of Silicone/Hydrogel Contact Lens......... 211
                    10.2.3.1   Advantages and Disadvantages of Silicone/
                               Hydrogel Contact Lenses 211

10.2.3.2   Industrial-Scale Batch and Continuous
Operation of MLCVD CH$_4$ Coating 216
10.2.4   Encapsulation of Metallic Stent 219
10.2.5   Plasma System Approach Interface Engineering for
Biomedical Electronics Devices 220
10.2.6   Unique Features of MLCVD Amorphous Carbon
Nanofilm for Biocompatibility 228
10.3   Interface Engineering for Adhesion of Coating 230
10.3.1   Salt Diffusion versus Salt Intrusion 230
10.3.2   System Approach Interface Engineering (SAIE) 236
10.3.2.1   Why Do We Need System Approach
Interface Engineering (SAIE)? 236
10.3.2.2   Conventional SAIE and Plasma SAIE
(P-SAIE) 237
References 239

Index 241

Preface

This book is an extension of the author's views, described in previous books, on magneto-plasma polymerization based on recent research on this subject. The previous books, *Plasma Polymerization* (Academic Press, 1985) and *Luminous Chemical Vapor Deposition and Interface Engineering* (CRC Press, 2004), describe the author's view on the fundamentals of low-pressure plasma polymerization. What triggered the author to write another book on the sections partly covered in the previous books were recent findings, which do not seem to be related to each other but seem to have important implications in regard to those issues.

The method described in this book, magneto-luminous chemical vapor deposition (MLCVD), is a perfect example of the "front-end approach green process," which utilizes an entirely new process that does not require any environmental remediation process. The method utilizes the minimum amount of materials in gas phase and yields virtually no effluent. The "rear-end green process" requires add-on processes to conventional chemical processes to take care of effluent problems.

The key process of dielectric breakdown of gas molecules under the influence of a magnetic field is a completely different phenomenon that has not been adequately described previously. It is also recognized that the distinction between molecular gases such as methane and trimethylsilane and mono-atomic gases such as helium and argon has not been seriously taken into account when dealing with the dielectric breakdown of the gas phase under low pressure. Nearly all fundamental investigations on this subject have been carried out with mono-atomic gas, and the knowledge gained from those studies has been intuitively assumed to be applicable to molecular gas. It has become convincingly clear that such an assumption should not be used when dealing with molecular gases that cause plasma polimerization.

It was also realized that the nanofilms prepared by MLCVD have unique imperturbable interfacial characteristics that were needed in the preparation of a surface that can be tolerated in various biological environments. Based on this realization, the author developed the minimum perturbation theory of biocompatibility, which can be considered as the neutral approach in dealing with biocompatibility and is at the opposite end of the biomimicking approach in pursuit of the creation of biocompatibility. The author felt it was his duty to explore the possibility of imparting biocompatibility to artificial materials by means of magneto-luminous chemical vapor deposition.

The final rather unusual and seemingly unrelated topic that encouraged the author to pursue writing this book was the National Aeronautics and Space Administration's (NASA) recent discovery that the magnetic field burst from the earth to space nearly half the distance to the moon triggers the inception of the aurora borealis (northern lights). The NASA Thimis project used four satellites placed three different distances from the earth two satellites on the lowest orbit. Although it has been known for nearly half a century that the aurora inception is tied to the magnetic field irregularity, this was the first unequivocal proof that the aurora starts when the magnetic burst from the earth occurs. What really intrigued the author was how it happens, and the NASA report seems to be rather vague. The author found that there is similarity in the inception of the magneto-luminous gas phase, which is used in the process described in this book, and the possible mechanism that starts the aurora. It is the author's belief that when we prove the dielectric breakdown of molecular gases such as nitrogen and oxygen under the influence of magnetic field and even one step further, the magnetic field inception of the luminous gas phase, we could solve both questions.

Histories of advancement of scientific knowledge clearly tell us that there is nothing more important than alternative views challenging or contradicting the prevailing concepts at the time and serious examination responding to the alternative view which leads to the modification of the concept approaching the truth. The truth is always beyond our perception of the truth, in the author's view. Albert Einstein stated that "imagination is more important than knowledge." What is really important is the imagination at the level of understanding. Without progressive imagination with advancing knowledge, or an open mind in pursuit of the truth, the knowledge could become an impediment in recognizing the revealing truth. This book contains numerous alternative views, and this could cause some difficulties in comprehending the contents; however, the author is confident that this book would stimulate readers to advancing scientific knowledge and developing new technologies.

The author is deeply indebted to many colleagues (coauthors of papers published on the related subjects) who have worked out experimental studies in the topical area addressed in this book with enthusiasm and dedication in pursuit of the truth. The author's special thanks are due to his wife, Gerda—without her understanding and sacrifices in daily life, this book could not have been completed.

The author would like to express his deep appreciation for the editor of this book series and the publication editors at Taylor & Francis, who accepted and encouraged the publication of this book in the green process series of books.

The Author

Hirotsugu Yasuda is an emeritus professor of chemical engineering at the University of Missouri, Columbia. He is a member of a 1960s pioneering group of plasma polymerization researchers and sole promoter of magneto-luminous chemical vapor deposition. He has written two books, *Plasma Polymerization* (Academic Press, 1985) and *Luminous Chemical Vapor Deposition and Interface Engineering* (CRC Press, 2004). He has also written 26 chapters in books in polymer science and engineering. He has published over 315 papers in refereed journals in the topical areas of plasma polymerization, gas permeability, biocompatibility, surface and interface, plasma interface engineering, polymer science, and applied polymer science. Yasuda has a unique combination of expertise in the three domains of science and technology: transport in polymer, biocompatibility, and plasma polymerization and has extensive experience in the biocompatibility of materials. He had continuous support from the National Institutes of Health for over 10 years in his research on the blood compatibility of surfaces prepared by plasma coating. The surfaces prepared by plasma coating were used to establish the concept of *imperturbable surface* in the athrombogenic approach, which relies upon a totally nonresponding surface.

1

Introduction

Chemistry has contributed to advancing quality of life as well as the comfort of the living. The positive contributions of industrial chemistry and biochemistry are numerous: infant mortality dropped, life expectancy increased, food production yield increased, and so forth. On the other hand, the negative side of the contribution should not be ignored. The prosperity of society and advancement of civilization have been achieved at the expense of environmental damage. Unless we minimize or totally eliminate the damaging side, we have to face the fact that our life cannot be sustained on a damaged planet earth, and advancements realized by chemistry and science and technology will eventually accelerate our fate.

The green processes are aimed at reversing the past trend and creating new processes that minimize or totally eliminate damaging effects on the environment. Green processes can be achieved by two fundamentally different approaches, the rear-end approach and the front-end approach. The rear-end approach is essentially the creation of additional environmental remediation processes that prevent the harmful byproducts of a chemical process from reaching the environment, by keeping them attached to the existing chemical process. The front-end approach is the creation of a new process that achieves the goal but does not yield the environment-damaging effluent. The latter is obviously the preferred approach, if we could create such a process that produces the aimed product with acceptable cost. The feasibility is dictated by economic factors at the time of development.

The cost of the environmental remediation process, on the other hand, was not accounted for in the cost of industrial operation for many years. Effluents were released in streams, oceans, or air, which caused serious damage to the environment. The cost of environmental remediation after industrial effluent is released is high, and is a very difficult and often impossible effort. Therefore, the effluent cleaning process must be added to the industrial process at the planning stage, which obviously increases costs. The decisive economic factor is the comparison of the overall costs of the two approaches—the cost of one new green process versus the overall cost of a non-green process plus an effluent cleaning process. The example described in this book clearly shows the importance of the front-end approach.

The magneto-luminous chemical vapor deposition (MLCVD) is a unique nanofilm deposition process, which is a dry process that does not use liquid chemical. There is practically no effluent, because it operates at very low pressure due to the unique influence of a magnetic field superimposed to the creation of the luminous gas phase, which cannot be reached by most vacuum deposition processes. The nanofilm deposition by MLCVD should be carried out as part of a well-designed interface engineering process, which requires thorough comprehension of the surfaces of various materials as well as what is to be accomplished ultimately by the process (i.e., there is no generic green process).

This book covers not only the detailed process of MLCVD but also basic principles involved in surface and interface. *Surface* is misleading terminology to describe *interface*, because every surface has the contacting medium—solid, liquid, gas, or even nothing (vacuum). The contacting medium influences surface characteristics of the material, which are dictated by the change of surface state that is the top surface region, roughly 10 to 40 nm of a solid phase. Molecular configuration of a monomer could change the bulk properties of a polymer, for example, but cannot change the surface characteristics as generally anticipated, because the surface characteristics are dictated by the surface configuration, which is determined by what moieties are actually at the interface. The surface configuration is dictated by the surrounding medium but not by the molecular configuration of the solid. The surface state of a polymer is generally highly perturbable by the contacting medium.

The significance of MLCVD is the creation of an imperturbable surface state and grafting onto the substrate material, yielding a material with an imperturbable surface state. It has been shown that the imperturbable surface state plays a crucial role in the biocompatibility of man-made materials in various situations encountered in biomedical applications. This principle is also applicable in corrosion protection, sustainable electric insulation of noble metals in biomedical applications.

In efforts to elucidate fundamental mechanisms of the MLCVD process, it has been convincingly evident that the luminous gas phase (plasma phase) of molecular gases created under the influence of a magnetic field significantly differs from that created without influence of a magnetic field. First, the initial steps of dielectric breakdown of a gas phase consisting of a monoatomic gas, such as He and Ar, and that consisting of a dissociable molecular gas, such as methane and trimethylsilane, are significantly different in that the knowledge gain in the former could not be applied to the latter. Furthermore, the fundamental mechanism of the dielectric breakdown of the gas phase is different because electrons in the magnetic field influence the electron-impact reactions with organic molecular gas (mainly methane described in this book) in a low-pressure domain that cannot be reached by most other vacuum deposition processes. It is interesting to note that the dielectric breakdown of molecular gas in a magnetic field provides an alternative mechanism for the inception of the aurora borealis (northern lights).

The recent NASA project to investigate the sudden magnetic field burst from the earth by placing five satellites in the same orbit at different altitudes found that the magnetic field burst and the appearance of aurora borealis occur concurrently. One critically important point, however, is that the aurora appears first as soon as the magnetic field burst takes place, contrary to the expectation that the aurora should appear at the end of the sequential happenings. This expectation seems to stem from the gas phase ionization concept. Although the correlation between aurora and the magnetic field burst has been established, the mechanism by which aurora develops seems to remain unsolved. Our recent reexamination of the dielectric breakdown of molecular gas (in low pressure) revealed that the fundamental mechanism of gas phase breakdown is significantly different from the classical interpretation of the phenomenon, which has been investigated mostly with nondeposition gas.

The emission of electrons from the cathode surface is caused by the applied electric field rather than the bombardment of ions on the cathode surface (i.e., the primary electron emission by electric field), and the gas phase breakdown under the superimposed magnetic field is significantly different from that without a magnetic field. It is observed that a magnetic field causes the inception of discharge under certain conditions, of which details warrant further investigation. Further examination of the gas phase breakdown mechanism under the influence of a magnetic field would cast light on how the aurora is induced by the magnetic field burst, which is low-energy excitation of molecular gas, oxygen, and nitrogen. The distinction between mono-atomic gas plasma and molecular gas plasma has been lacking in the conventional interpretation of gas phase breakdown. Furthermore, the distinction between deposition plasma and nondeposition but reactive molecular gas plasma has not been adequately dealt with in polymerization or plasma deposition of organic molecular gases, but plays a key role in understanding and perfecting the plasma polymerization process.

Plasma interface engineering under the scheme of system approach interface engineering is the front-end approach green process. The thickness of interfacial layers applied is at least three orders of magnitude smaller than conventional coating applications to achieve the same effect; however, adhesion improvement and barrier characteristics achievable by plasma interface engineering are at levels that cannot be obtained by a conventional coating approach, which could be a rear-end green process, at best. For delicate biomedical device applications, plasma interface engineering is the only applicable technology. MLCVD (magnetron plasma polymerization) has the capability of performing such delicate operations in larger-scale continuous operation in the mode of *front-end super green process*.

2

Context of Terms Used and Concepts

Subjects addressed in this book encompass numerous cross-disciplinary academic topics and application domains. The main interest and understanding of electrons, for instance, differs depending on the discipline; physicists are mainly concerned with core-level electrons, chemists with valence-level electrons, and biologists with electrons in free radicals. Likewise, terms customarily used in one discipline are also used in different disciplines, but often in significantly different contexts.

Shortly after the 9/11 (2001) attacks, the author, while at the Surface Science and Plasma Technology Center, University of Missouri, received a telephone call from the local Red Cross office offering blood plasma for research without charge, because they received an overwhelming number of blood donations that created a huge surplus of blood plasma, which should be used within a certain period of time. It was necessary to explain the different meaning of *plasma* in order to express our thanks for their kind offer and decline it. This example clearly explains why it is necessary to have this section appear early in this book.

Terminologies described below are not generic definitions of terms but are the contexts of terms and some concepts used in this book, which are necessary to comprehend the content of this book without raising unnecessary questions or arguments.

2.1 Plasma

A more comprehensive explanation of the term *plasma* is given in Chapter 3. The term *plasma* is used to identify the gas phase that glows in low-pressure electrical discharge of gas without further identification of component species.

2.1.1 Equilibrium and Nonequilibrium Plasma

The plasma phase that retains the energy equilibrium of component species is described as equilibrium plasma, and that which does not is described as nonequilibrium plasma. Low-temperature plasma or low-pressure plasma is nonequilibrium plasma. Equilibrium plasma exists only in high-temperature plasma, which is synonymous with thermal plasma, plasma created by thermal energy. Electroneutrality, when the number of electrons and ions are equal, can be maintained only in equilibrium plasma.

2.1.2 Low-Pressure (Low-Temperature) Plasma

These terms are used to point out that the plasma state referred to is not high-temperature or equilibrium plasma. Low-pressure plasma is created by electron impact reactions of gas, and electroneutrality cannot be intuitively assumed.

2.1.3 Deposition Plasma and Nondeposition Plasma

A plasma phase that causes deposition of the condensed phase material of the gas used is termed *deposition plasma*, which depends entirely on what kind of gas is used to create the plasma state. The plasma phase that does not yield the deposition of material is collectively termed *nondeposition plasma*.

2.1.4 Reactive Plasma and Nonreactive Plasma

Nondeposition plasmas are further divided into reactive plasma and nonreactive plasma depending on whether or not gas used chemically reacts with solid that contacts the plasma phase (e.g., oxygen plasma is a reactive plasma, and argon plasma is a nonreactive plasma). Gas in reactive plasma is consumed by chemical reactions. Gas in nonreactive plasma is not consumed but transfers energy to materials that contact the plasma (substrate), which could cause a physical or chemical reaction within the substrate.

2.1.5 Plasma Sheath

The low-pressure gas phase has no discernible surface state due to the high mobility of gas. The gas phase is uniform up to the interface with a solid wall. However, there are identifiable surface states of plasma, termed *plasma sheaths*. A plasma sheath is the energy-transferring zone in the plasma phase near the contacting surface. Accordingly, there are two characteristically different plasma sheaths—the energy input plasma sheath and the energy dissipating plasma sheath. The characteristics of a plasma sheath depend on the type of gas, the energy input, pressure, and other electrical discharge conditions. The characters of the plasma sheath of low-pressure plasma depend on the mode of electric power input to the gas phase.

In low-pressure plasma, the sheath is not due to separation of electron and ion, because the electroneutrality does not exist in the plasma phase created by the electron impact reaction of gas.

2.1.6 Magneto-Plasma

Plasma is created under the influence of a magnetic field in the gas phase. The gas phase breakdown mechanism is significantly different than that without a magnetic field.

2.2 Electric Discharge of Gas

2.2.1 Direct Current (DC) Discharge

Electric discharge created by a direct current (DC) power source is termed *DC discharge*. Discharge voltage and current are directly measured, and discharge power is calculated from those values. Discharge current is the most important operation parameter. Electrons in linear motion cause electron impact reactions with gas near the surface of the cathode.

2.2.2 Audio Frequency (AF) Discharge

Electric discharge created by a high-frequency power source (e.g., 10 to 40 kHz) is referred to as *audio frequency (AF) discharge*. Frequency does not refer to the audible frequency. AF discharge is essentially alternating polarity DC discharge.

2.2.3 Radio Frequency (RF) Discharge

Electric discharge created by a radio frequency power source (e.g., 13.5 MHz) is referred to as radio frequency (RF) discharge. RF discharge is significantly different from DC and AF discharges in its dielectric gas phase breakdown mechanism. Discharge power is the operation parameter. Electrons in oscillating motion cause an electron impact reaction of gas.

2.2.4 Cathode Fall

In DC discharge of an atomic gas, there is a dark space between the cathode surface and the glow, which is called the *cathode fall* region. Electrons are accelerated by the electric field in the dark space. In RF discharge, there is also a dark space between an electrode (if it was used) and the glow, but it does not have the same function as the cathode fall of DC discharge, because the motion of the electrons is completely different (i.e., oscillating electrons).

2.2.5 Ionization Glow

The glow created by the ionization of gas appears at the end of the cathode fall region in DC discharge of nondissociable gases.

2.2.6 Dissociation Glow

The glow created by the electron-impact dissociation of molecular gas appears in DC discharge as cathode glow that virtually touches the cathode surface.

2.2.7 Self-Bias and Bias Potential

The differential potential observed between the powered and grounded electrodes, mainly in RF discharge, is termed *self-bias potential*. Self-bias generally is due to the asymmetry of the electrode surface area, because any grounded metal surface existing in a reactor functions as the grounded electrode. Bias potential generally refers to the added DC potential to an electrode or conducting substrate holder or substrate. The superimposed DC discharge is generally ignored.

2.2.8 Deposition Gas

The gas used in electric discharge which causes the deposition of materials by itself (i.e., without the addition of another gas) is referred to as *deposition gas*. Deposition gases are dissociable molecular gases.

2.2.9 Nondeposition Gas

Gas that does not cause deposition of material by itself is referred to as *nondeposition gas*, which could be a nonreactive gas such as argon or a reactive gas such as oxygen.

2.2.10 Electronegativity

Electronegativity of an atom is the affinity of the electron to the atom. High electronegativity attracts electrons, and low electronegativity repels electrons. The electronegativity of an atom within a molecule greatly influences the electron impact reaction of the molecule. In other words, an electron does not hit a molecule in a random mode, but the location of impact is guided by the electron negativity of the atom within a molecule.

2.3 Gas Phase

2.3.1 Gas Equation

The gas law that defines the gas phase is given by $pV = nRT$, where p is the pressure, V is the volume of the system, n is the number of gas atoms or molecules, R is the gas constant, and T is the temperature of the system in Kelvin.

The plasma phase is no exception to this rule. The number of gas molecules determines the pressure of the system, but mass deposited from the plasma phase depends on the mass of species involved in the plasma phase which deposit as condensed phase materials, which is also governed by the gas law.

The pressure in a flow system depends on the difference in the gas feed-in rate and the rate of pumping out. Dealing with nonreactive, nondeposition gases such as argon or helium, the dynamically maintained pressure and the pressure defined by the gas law are the same for all practical purposes. However, dealing with consuming gas and deposition gas, the dynamical system pressure depends on the rate of consumption of gas due to chemical reaction and the rate of production of nonreactive gases in addition to the gas flow rate factors. Consequently, the actual pressure of plasma depends on how the dynamic pressure is maintained.

2.3.2 Flow Rate, Volume Flow Rate, Mass Flow Rate

Because of the gas law, the flow rate of gas is generally given by the volume flow rate (e.g., cc (STP)/m, which is the flow rate of a number of gases). Any parameter that defines a chemical reaction, on the other hand, depends on the unit of mass (e.g., reaction energy is given by J/mole), of which mole is the mass of 1 mole (number) of gas molecules (i.e., J/kg).

2.3.3 Ionized Gas

Ionized gas means the majority of gaseous species are ions. A gas phase in which ionized gas is less than, for example, 10^{-2}, is not appropriate to be identified as ionized gas.

2.3.4 Luminous Gas

The gas phase that has luminosity, or glow, is expressed by luminous gas. Luminous gas (phase) is used synonymously with plasma (phase).

2.3.5 Photo Emitting Excited Neutrals

The major component species in a luminous gas phase created in low pressure are excited neutral species that emit photons.

2.3.6 Magnetic Field in Gas Phase

Magnetic fields extended into the gas phase from a magnet consist of a number of field lines, along which electrons circulate and move in the direction of the magnetic field in a spiral motion. The extension of a magnetic field from a magnet to gas phase occurs independent of the type of gas and the number of gases in the gas phase (i.e., independent of pressure).

It is considered that electrons in the magnetic field cannot resist colliding with gas molecules in the gas phase. The consequence of an electron–molecule collision is dependent on the energy level of the electron in the magnetic field, which is proportional to magnetic field strength, and the overall number of collisions per molecule. The collision of electrons in a magnetic field with molecules in the gas phase dramatically changes the fundamental gas phase breakdown process.

2.4 Polymerization and Material Formation

2.4.1 Plasma Polymerization

Formation and deposition of materials from small molecular weight molecules in the gas phase by electric discharge, usually in low pressure, is termed *plasma polymerization*, because the entire process occurs in the plasma phase.

2.4.2 Chemical Vapor Deposition (CVD)

Chemical vapor deposition (CVD) is developed as a process that creates solid phase materials from gaseous inorganic materials in the gas phase, usually at elevated temperature (e.g., > 600 K). It is a thermally driven deposition process.

2.4.3 Plasma Chemical Vapor Deposition (PCVD)

This process is plasma-driven chemical vapor deposition, which can be used synonymously with plasma polymerization but with the main emphasis on the processing. PCVD is not a thermally driven process but a plasma-driven process.

2.4.4 Plasma Enhanced CVD (PECVD); Plasma Assisted CVD (PACVD)

The original use of the terms *plasma enhanced CVD* and *plasma assisted CVD* literally meant to reduce the temperature of the CVD process, and they are still used in this context. Those terms are also used to describe the processes that are practically identical to plasma polymerization. A common denominator factor with –CVD processes is that less or no emphasis is placed on chemical reactions involved in the process.

2.4.5 Luminous Chemical Vapor Deposition (LCVD)

The term *luminous CVD* is used synonymously with plasma polymerization with the concepts that the luminous gas phase used is not an ionized gas phase, and the reaction mechanisms are not conventional in the sense of polymerization.

2.4.6 Magnetron Plasma Sputter Deposition

Magnetically enhanced plasma is used to sputter metal used as the cathode of DC discharge and deposited on the substrate placed away from the cathode. Magnetron plasma polymerization or magneto-luminous CVD uses the similar plasma generation principle but suppresses the sputtering of metal by selecting the gas, pressure, and magnetic field strength.

2.4.7 Magneto-Luminous CVD, Magneto-Plasma CVD, Magnetron-Plasma Polymerization

Those terms are used synonymously with the ideas described above.

2.4.8 Material Formation Mechanism

Mechanisms of material formation by plasma polymerization are complex as described in Chapter 7. One important feature is that the gas used for the process is not the species that contributes material formation, but species created by the electron-impact dissociation of the original gas are involved in the material formation.

2.4.9 Deposition Mechanism

In the overall process of material deposition by plasma polymerization, the process of creating reactive species and the process of deposition are not coupled. Because of this feature, significantly different material deposits in an otherwise identical process by simply changing the temperature of the substrate (change of deposition condition).

2.5 Surface and Interface

2.5.1 Surface

Surface is the end of a condensed phase material usually in contact with ambient air, which is actually, in most cases, the interface of said condensed phase and the ambient air.

2.5.2 Interface

Interface is the boundary of two dissimilar materials. Every "surface" is an interface with the contacting medium. Interfacial properties, therefore, cannot be defined without identifying the contacting medium.

2.5.3 Interfacial Electron Transfer

When contact is made between two different surfaces, the electrons in each surface state migrate from one surface state to another surface state in contact according to the difference of the overall electron affinity. The electron transfer occurs without an external driving force. This sets the fundamental principle for how electrons move from one surface (e.g., electrode surface) to the contacting gas phase under electric voltage difference (i.e., the electron emission from the cathode surface does not require the high energy conventionally assumed).

2.5.4 Surface Static Charge

Static charge is created on the surface of a dielectric material when two materials in contact are abruptly separated, because the gain or the loss of electrons on the contact remain on the surface. This also provides an example of characteristics of electron movement across different surfaces.

2.5.5 Molecular Configuration

Molecular configuration refers to the arrangement of atoms and chemical moieties within a molecule.

2.5.6 Surface Configuration

Surface configuration refers to which atoms or moieties are at the surface. The surface configuration determines the surface property, not the molecular configuration. The surface of hydrophilic material based on the molecular configuration could be, and often is, hydrophobic. This sets the goal for how plasma surface modification should be performed depending on the objective.

2.5.7 Surface State

The thin top layer (e.g., 20 nm) of the material is significantly different from the bulk phase of the material because the force balance maintained within the bulk phase cannot be maintained in the top layer. This thin top layer is termed the *surface state* to distinguish it from the bulk phase state. The details of a surface state usually change depending on what contacts the surface state.

2.5.8 Imperturbable Surface State

The imperturbable surface state is the stable surface state that cannot be perturbed by the difference of the contacting medium. The surface state of gold or platinum is an example of an imperturbable surface state, however, with molecular materials; imperturbability of a surface is synonymous with imperturbability of surface configuration.

2.6 Biocompatibility

2.6.1 Biocompatibility of Artificial Material

The term *biocompatibility* is not a scientific terminology, because *biocompatibility* of a material depends on the host biological system. However, this term is widely used to grade or claim acceptability of a material in a biological system. Biocompatibility in a generic sense cannot be considered; one material in a specific form could be compatible in a specific part of the body but may not be compatible in different parts of the body.

2.6.2 Sustainability of Biocompatibility

Sustainability of biocompatibility (in a specific set of conditions) depends on how the surface state could be maintained unchanged for a prolonged period of time, which more or less requires the imputable surface state described above.

2.6.3 Minimum Perturbation Theory of Biocompatibility

Biocompatibility of an artificial material in a specific biological system depends on the size and shape of the overall article to be placed in the host biological system. If the overall perturbation caused to the host biological system by the foreign body is below the threshold level that can be tolerated by the host biological system, the foreign body could remain in the host biological system; it is based on the neutrality or the nonreactivity of the surface.

3

Green Deposition Coating of Nanofilms

3.1 Front-End Approach and Rear-End Approach in Green Processing

Dealing with the objective of green processing in general, there are two major distinctively different approaches we could take: one is the *front-end approach*, and the other is the *rear-end approach*. In the front-end approach, we develop a new process that yields the minimum effluent and does not require an elaborate environmental remediation process, but it is costly. Obviously, this approach is not applicable to any chemical reaction process except in very special cases, one of which is described in this book as magneto-luminous chemical vapor deposition (MLCVD). In the rear-end approach, chemical processes are carried out with minimum consideration of the *green* aspects of the processes, but an elaborate environment remediation process is developed to take care of the overall greenness of the chemical process. In this context, environmental remediation does not mean the cleaning of a polluted environment but rather the special treatment process of effluent before releasing to the environment. In many chemical reactions, the rear-end approach is the only plausible method one could take.

The front-end approach green process requires development or creation of an entirely new approach, which in most cases costs more than ongoing processes. The economical factor of a processing, however, should be examined considering the total cost of the process as a whole, but not by any specific item such as the initial cost of the equipment alone. The cost of the environment remediation process accounts for a considerable portion of the total cost. An important factor changing the equation for the economy of chemical processing in recent years is the cost associated with the requirement to keep the environment clean and to minimize health hazards to the workers involved. In other words, the aspect of *green processing* has become the most important factor in determining the economical acceptability of a chemical processing.

How a low-pressure MLCVD coating fares with a conventional wet chemical coating is examined in the case of nanofilm coatings applied on a contact lens. Figure 3.1 shows a pictorial view of a contact lens on which a very thin layer (ca 20 nm thick) of coating is applied. Coating is applied to modify

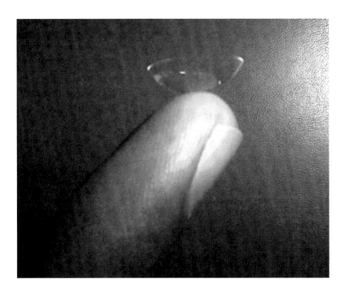

FIGURE 3.1
A contact lens that will be coated by two different processes.

surface characteristics (both sides) of the contact lens, which is necessary to make the special types of contact lens wearable and improve the other functional characteristics. The details of coating processes and characteristics of coating are described in Chapter 10 (Section 10.2.3), and only operation costs of two fundamentally different processes are compared in this section.

3.1.1 Low-Pressure Magneto-Luminous Chemical Vapor Deposition (MLCVD) Coating

Contact lenses are coated with an approximately 20 nm thick layer of an amorphous carbon film on both sides of the lens by a continuous linear mode operation, which is schematically shown in Figure 3.2. The total processing time for a contact lens is approximately 40 minutes, which includes drying of the contact lenses, evacuation of samples, MLCVD coating, and repressurizing to ambient atmosphere. The coating operation is continuous for approximately 30 days between maintenance breaks. The reactor is capable of coating 30 million contact lenses a year (340 days of operation).

3.1.2 Layer-By-Layer (LBL) Coating

The same contact lenses are coated with an approximately 20-nm layer, consisting of alternating layers of an acidic polymer and a basic polymer, by consecutive layer-by-layer application (LBL coating) of respective dilute solutions of the two types of polymers, which represents a case of rear-end approach green process because the process requires effluent (large quantities

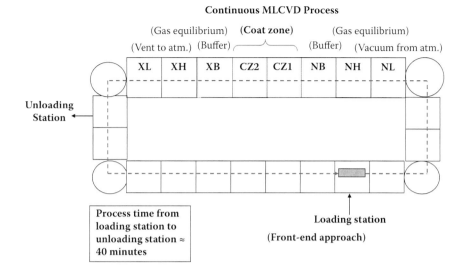

FIGURE 3.2
MLCVD contact lens coating line.

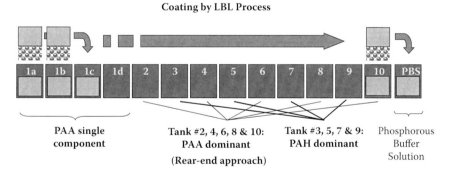

FIGURE 3.3
An LBL coating line.

of leftover dilute solutions) treatment. The LBL coating process is schematically shown in Figure 3.3. Ten baths are placed in a line. A sample holder with multiple contact lenses is immersed in the first bath. The technical features of LBL coating and some properties of coated lens are also described in Chapter 10.

The first application of acidic polymer solution is repeated three times. The first coating is the most difficult, because it is essentially trying to coat a very hydrophobic surface with an aqueous solution of ionic polymer, and this is the most serious drawback of the coating applied on a silicone/hydrogel contact lens. In order to improve the chance of establishing the foundation for subsequent steps of alternative coatings, the first-layer coating is repeated three times. Then the first layer of basic polymer solution is applied, which reacts with the coated layer of acidic polymer to form a polymer–polymer salt. The repeated alternative coatings of acidic polymer and basic polymer make an excellent layer of highly hydrated hydrogel film.

The removal of excess solution is important in minimizing the reaction of acidic polymer solution and basic polymer in the solution of the subsequent bath containing polymer solution of opposite polarity. Due to cross-contamination of the two kinds of polymer solutions, the solution baths should be cleaned, and new sets of dilute solutions should be used after 5 to 10 batches of coating operation. The calculation of the processing cost shown below is based on 10 batches before the change of solutions. The total capacity of the coating process is determined by the number of lines to be employed. The time necessary to complete coating is roughly the same with that for a MLCVD coating without counting the time necessary to change solutions, but the processing is slightly more labor intensive than the highly automated MLCVD line operation within a continuous operation cycle.

The most crucial factor, as far as the cost estimate is concerned, is the number (x) of coatings that a bath containing a type of polymer solution could be used for without changing the solutions. If the value of x decreases below 10, the material cost increases proportional to $10/(10 - x)$; that is, if the solutions must be changed after dipping five batches, the material cost will increase to twice the cost cited in the cost estimate.

3.1.3 Cost of Coating Processes

The cost of the two coating processes for production of 28 million coated contact lenses is compared in Figure 3.4 [1]. The cost of the MLCVD coating system is roughly 12 times more expensive than that for the LBL coating system. However, the plasma coating is an ultimate green process that uses the minimum amount of materials, and there is no effluent in a practical sense. Only 11 kg of coating materials (methane and air) is used to coat 28 million contact lenses in a year. The coating yield (amount of material deposit on contact lens/total amount of materials used) is roughly 30% with the reactor used for a cost estimate. The cost of coating materials is only $140 for 1 year's operation. Because there is no effluent, there is no cost for effluent treatment. The labor cost is not included in the table; however, it is estimated that LBL coating is roughly 40% more labor intensive.

The LBL coating process requires the use of very dilute solutions of polymers in order to coat a layer with thickness in the range of 1 to 2 nm on each dip-coating operation. This means that a large amount of solvent (water) is needed to make the coating solutions very dilute. The yield of nanofilm coating by layer-by-layer application of very dilute solution is roughly 0.05%—that is, most expensive polymers end up in the effluent. The amount of coating materials (polymers and water), of which the majority is water, needed for

	Rear End Approach		Front End Approach	
	LBL Coating		LCVD Coating	
Initial Cost of Coater	$250,000		$3,000,000	
Annual Operation Cost				
	Quantity (Kg)	Cost (US$)	Quantity (Kg)	Cost (US$)
Coating Materials	7,759,257.4	$181,277	1.9	$146
Coating Yield	0.05%		13.59%	
Total Effluent of Processes	374,400		~ 0.2	
Effluent Treatment Cost		$5,971,292		$0
Depreciation		$25,000		$300,000
Total Cost (without labor)		$6,177,569		$300,146

FIGURE 3.4 (*See color insert.*)
Comparison of operation costs for conventional wet chemistry process and LCVD coating and impact of environmental remediation cost.

the production of 28 million contact lenses is roughly 7,760,000 kg, and its cost is roughly $180,000 in contrast to $146,000 for the MLCVD coating.

If we could neglect the cost of wastewater treatment, labor cost to operate, and additional labor cost of changing solutions, the overall cost of the coating processes is roughly the same (LBL is two-thirds the MLCVD coating), in spite of the fact that the cost of coater for the plasma coating is 12 times more expensive than equipment needed for the LBL coating. If a solution cannot be used for 10 coatings, the cost for the LBL coating (without wastewater treatment) will become greater than that for the MLCVD coating. If, for example, the coating solution should be changed after five coating runs, the cost of materials as well as the cost of wastewater treatment will be twice as much as shown in Figure 3.3. The change of polymer solutions is dictated by the quality of coating applied on contact lenses and cannot be manipulated for the sake of the operation cost. Considerable cross-contamination is caused by the solution carried by the lens-holding devices and other sample transporting mechanisms.

With the cost for wastewater treatment, plasma coating is a much more economical coating process than the supposedly inexpensive LBL coating process. The cost of wastewater treatment included in Figure 3.3 is based on the commercial wastewater treatment cost of $1,000 for a 55-gallon drum. With this wastewater treatment cost, the overall cost of the LBL coating operation is 11.4 times more expensive than that of the MLCVD coating process. It should be recognized that the cost of wastewater treatment for 1 year of operation, which is a recurring cost, shown in Figure 3.4, is nearly twice the original cost of the MLCVD reactor. This example clearly demonstrates how wrong the judgment based on only the original cost of vacuum equipment could be. Other factors that should be considered are that water is not free and its cost is rising rapidly, and that effluent, dilute aqueous solution, should not be dumped into rivers or seas.

The cost of wastewater treatment for 1 year's operation by LBL coating is more than twice the original cost of the plasma coating system, which is often conceived as prohibitively expensive for industrial applications. The "green" aspect of processing is becoming increasingly important, and the cost for green processing might become the major factor, as shown in this chapter.

Vacuum processing, in general, could be the ultimate in green processing in that the minimum quantity of materials are consumed and the effluent is contained in the system, which makes the effluent treatment, if necessary, easily accomplished. So far as green processing is concerned, vacuum processing is a typical example of the front-end approach (i.e., creating or choosing processing that does not pollute the environment). The LBL coating process is a typical example of the rear-end approach, which was previously viewed as an inexpensive process (i.e., carrying out processing without considering the environmental influence in advance), with the environmental pollution being taken care of by the subsequent remediation process. This process is more difficult and is often prohibitively expensive and will be more so in the future.

The data presented here clearly show the following important aspects of vacuum processing, which are often not recognized or ignored:

1. Vacuum processing could be an ultimate green processing, if we choose an appropriate process that satisfies the necessary requirements to be an ultimate green process.
2. The cost of vacuum equipment could be offset by the favorable cost of operation that uses the minimum amount of materials and does not require the treatment of effluent or practically no effluent.

The MLCVD process, uniquely qualified to be an ultimately green process, and its fundamental principles are explained in detail in the following chapters. However, it should be noted that not all low-pressure plasma deposition processes could be considered to belong to this category of green processing. In some processes, expensive and rather toxic gases are used, employing higher pressure and higher flow rate than the MLCVD process and hence cannot be considered as green processes without extensive effluent recovery and possibly detoxification processes.

It is often considered that low-pressure plasma processes are expensive simply because of the high cost of vacuum operation, and it is advantageous to use processes that do not use vacuum. In many cases, however, the compensatory negative factors of the atmospheric pressure or superatmospheric pressure electric discharge process, including much higher gas consumption (2,000 times or more), higher electric power consumption, the necessity of having a gas recovery system, and so forth, are not considered for a similar-scale operation. Even in a simple case of atmospheric electrical discharge of N_2 gas in a similar-scale operation, the effluent gas cannot be released into ambient air in a room. N_2 is a typical nontoxic inert gas used in chemistry laboratories, but one should bear in mind the overwhelming effect of N_2-rich (O_2-depleted) atmosphere (i.e., N_2 is nontoxic but not life supporting).

The overall costs involved in the front-end approach green process and those for the rear-end approach green process are compared in this section. The more crucial issue, however, is the quality of coatings that could be provided to the surface of contact lens. In this requirement, MLCVD coating excels beyond comparison, as described in Chapter 10.

3.2 System Approach Interface Engineering with Green Processes

3.2.1 System Approach Interface Engineering for Corrosion Protection

Corrosion protection of metals, such as steel in automobiles and aluminum alloys in aircrafts, has been practiced by applying the concept of system approach interface engineering (SAIE), which means the control of interfaces in applications of various primer coatings considering the whole combination of various steps as a system. Conversely, it also means that mere combination of two or three best components does not yield better corrosion protection of the base metal. The practice of SAIE advanced the corrosion protection of metals, in general, at a remarkable level in the past decades (e.g., modern automobiles last much longer without corrosion than they did decades ago). However, on the other hand, SAIE are carried out by using chemicals that are hazardous and environmentally damaging (i.e., rear-end approach SAIE). Front-end approach green SAIE could improve the corrosion protection and eliminate problems associated with environmental remediation [2–9].

The initial cost of an LCVD reactor, which is substantial, is an extremely important factor in consideration of a new manufacturing operation for corrosion protection. Consequently, the feasibility of LCVD as a manufacturing process largely depends on the commercial value of the product and the add-on value attributable to LCVD. However, the decisive add-on value is an intangible value that depends entirely on the concept of the value of the product. If environmental greenness was considered very important, the front-end green process of plasma SAIE (P-SAIE) would be an ideal way to solve problems, by improving corrosion protection, keeping the environment clean, and more importantly, keeping workers from health hazards.

Major hurdles to overcome in the development and utilization of LCVD (plasma polymerization) processes for industrial-scale applications, outside of microelectronics applications, are nonfamiliarity of vacuum processing and resistance due to psychological fear of the unknown, and the relatively high initial cost of the equipment. In many cases, these two factors are strong enough to force planners to shy away from the potential use of low-pressure processing even before trying in laboratory-scale experiments, although the unique characteristics that can be obtained by LCVD have been known for many years. These two factors are not technical problems, and the second factor is strictly the poor judgment of the economical factor. The concept of green processes to protect our environment has been almost completely missing in the planning of new chemical processes in the past.

Practical use of P-SAIE could be examined with the corrosion protection of aluminum alloys for aircraft. In this case, the obvious advantage is the elimination of existing polluting and hazardous materials in the processing steps and also in the products. However, the greatest benefit of making a process environmentally benign is intangible. Keeping this great intangible benefit in mind, it is possible to compare tangible advantages in the cost of materials involved and the possible reduction of steps in the overall manufacturing process. As a transitional step, the LCVD process can be added by piggybacking on the ion vapor deposition (IVD) process, which has been used for aluminum alloy corrosion protection for military aircrafts. IVD is a deposition of inorganic material (aluminum oxide) in the DC discharge of argon. This piggybacking means that there is no need for separate equipment for LCVD. Figure 3.5 compares steps involved in the current IVD process and the IVD/LCVD hybrid process.

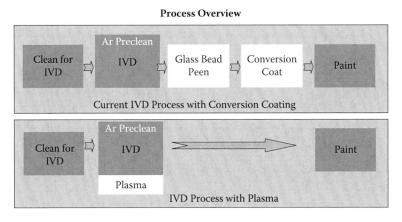

FIGURE 3.5
Comparison of IVD and IVD/LCVD hybrid processes.

LCVD adds the operation time of an IVD reactor but eliminates two subsequent processes (i.e., glass bead peening and chromate conversion coating) that have a processing time much longer than the added LCVD processing time. Consequently, the addition of LCVD reduces the overall process cycle time and labor cost associated with these two processes to be eliminated and the transferring process of substrates, which requires labor and time. Furthermore, not only is this process environmentally benign, but it also improves the corrosion resistance of a primer applied over it which does not contain environmentally hazardous chromate. All these are significant but rather intangible advantages in view of an ongoing process.

The largest cost that can be eliminated is the chromate conversion coating, and the cost for treatment of spent solution and rinse water (details of costs are not available for comparison). In comparison to these costs, the cost for trimethylsilane (LCVD gas) used in the closed-system LCVD operation is almost negligible. Keeping hazardous materials out of the overall processing system is very important in that workers are not exposed to health hazards in their environment, which is also an intangible benefit as far as cost comparison is concerned. (Due to the nature of processing, no cost figures are available to estimate cost savings by the front-end green process P-SAIE.)

There is a high possibility that the deposition of aluminum oxide could be totally replaced with LCVD of trimethylsilane which would further reduce the operation cost. If MLCVD could be used, instead of LCVD, the process could be operated in a continuous mode, though it requires a new reaction system, not piggybacking on the existing operation.

The major concern of "green processing" today is the green aspect of the production processes as described above. However, another important aspect of "green processing" is the influence on the recovery processes of the products produced by the "green processing." For instance, the products produced by rear-end approach SAIE (e.g., steel body of automobile or aluminum alloys of aircrafts) should go through the recycling process to recover steel or aluminum alloys some years later. Those metals contain metals that function in the galvanic corrosion protection, which works basically on the same fundamental principle of SAIE (i.e., they must be in the top layer that constitutes the interface with the environment but not mixed in the bulk phase). Zinc at the top layer of steel works well in protecting the underlying steel bulk phase, but if Zn is mixed uniformly in steel, it makes a useless alloy. Therefore, a dezincking process is necessary before recycling the steel. The same situation exists in Cr-containing alloy. Because of this requirement, a steel company stopped receiving the leftover zinc-coated steel from the stamping operation at car manufacturers.

The front-end P-SAIE has unique advantages in this respect. First, the corrosion protection does not depend on electrochemical, galvanic, corrosion protection, which requires a small amount of water at the interface in order to make the electrochemical reaction function. Second, but probably more importantly, we could choose elements that do not cause difficulty in the recycling process of the products. Elements so far examined are C and Si, which seem to be acceptable in recycling processes. Third, the amount of elements involved in the interfacial layer is roughly three orders of magnitude less than those in rear-end SAIE, which makes the recycling of metals more feasible.

3.2.2 Green System Approach Interface Engineering by Magneto-Luminous CVD

Magneto-luminous CVD (magnetron plasma polymerization) has been used extensively in the author's laboratories for over 35 years; however, the method has not been known widely beyond limited users of the method. Probably the method has been conceived as one of glow discharge polymerization, which is a little more complicated than equipment used in the laboratory, by most users of radio frequency (RF) plasma processes. The unique advantages of the method have become clear in relatively recent years by examination of the fundamental steps of the dielectric breakdown processes of gases in DC discharge with and without magnetic field, which is described in some detail in this book.

MLCVD operates with audio frequency (AF) power sources in the typical range of 10 to 50 kHz. Because of a unique mode of gas phase breakdown, the process can be operated in the low-pressure range, which is beyond the minimum pressure range obtainable by other modes of glow discharge, and makes the process super green. Low pressure means a lesser number of gases (i.e., the minimum amount of gas is used with practically no effluent from the reactor). This method has been utilized in some green P-SAIE with relatively small objects such as contact lenses, metallic stents, and bioengineering devices. Because of unique material formation mechanisms, the method has great potential to be developed for larger-scale production of larger-size products via P-SAIE as nearly total green processes, if the need arises. One important objective of this book is to provide information for finding needs for such a promising process, described as a supergreen process, of MLCVD coating of nanofilms.

References

1. Yasuda H. and Y. Matsuzawa, *Plasma Processes and Polymers*, 2, 507, 2005.
2. Wang, T. F., H. Yasuda, T. J. Lin, and J. A. Antonelli, *Progress in Organic Coatings*, 28, 291, 1996.
3. Chun, H. J., D. L. Cho, T. J. Lin, H. Yasuda, D. J. Yang, and J. A. Antonelli, *Corrosion*, 52(8), 584, 1996.

4. Yasuda, H., T. F. Wang, D. L. Cho, T. J. Lin, and J. A. Antonelli, *Progress in Organic Coatings*, 30, 31, 1997.
5. Lin, T. J., J. Antonelli, D. J. Yang, H. K. Yasuda, and F. T. Wang, *Progress in Organic Coatings*, 31, 351–361, 1997.
6. Reddy, C. M., Q. S. Yu, C. E. Moffitt, D. M. Wieliczka, R. Johnson, J. E. Deffeyes, and H. K. Yasuda, *Corrosion*, 56, 819, 2000.
7. Yu, Q. S., C. M. Reddy, C. E. Moffitt, D. M. Wieliczka, R. Johnson, J. E. Deffeyes, and H. K. Yasuda, *Corrosion*, 56, 887, 2000.
8. Moffitt, C. E., C. M. Reddy, Q. S. Yu, D. M. Wieliczka, R. Johnson, J. E. Deffeyes, and H. K. Yasuda, *Corrosion*, 56, 1032, 2000.
9. Yu, Q. S., C. M. Reddy, C. E. Moffitt, D. M. Wieliczka, R. Johnson, J. E. Deffeyes, and H. K. Yasuda, *Corrosion*, 57(9), 802, 2001.

4

Plasma Phase and Luminous Gas Phase

4.1 Plasma Phase

This state of matter was first identified by William Crookes in a Crookes tube and was described as "radiant matter" in 1879 [1]. The nature of the radiant cathode ray matter was subsequently identified by J. J. Thompson in 1897 [2], but the term to describe *plasma* was coined by Irving Langmuir in 1928 [3]. However, it is not clear why Langmuir picked *plasma*, though loose connection to "blood plasma" was mentioned [4].

The following explanatory excerpt is from G. L. Rogoff [15] p. 989:

> During the 1920s Irving Langmuir was studying various types of mercury-vapor discharges and he noticed similarities in their structure—near the boundaries as well as in the main body of the discharge. While the region immediately adjacent to a wall or electrode was already called a "sheath," there was no name for the quasi-neutral stuff filling most of the discharge space. He decided to call it "plasma."
>
> While his relating the term to blood plasma has been acknowledged by colleagues who worked with him at the General Electric Research Laboratory [4,5], the basis for that connection is unclear. One version [5] of the story has it that the similarity was in carrying particles, while another account [6] speculated that it was in the Greek origin of the term, meaning "to mold," since the glowing discharge usually molded itself to the shape of its container. In any case, it appears that the first published use of the term was in Langmuir's "Oscillations in Ionized Gases," published in 1928 in the *Proceedings of the National Academy of Sciences* [3].
>
> Thus the term "plasma" was first used to describe partially (if not weakly) ionized gases. The term plasma apparently did not find immediate widespread use in the scientific community. It did eventually catch on, however, but in some cases the term was inappropriately limited to highly ionized gases.

The use of the term *plasma* was widely used later in the study of high-temperature *plasma*, with strong emphasis on the ionized gas phase that retains electroneutrality (i.e., number of ions is equal to the number of electrons). Subsequently, at the dawn of interest in plasma surface modification in the early 1960s, the term *plasma* became the starting point of the technology development. The concept that plasma is ionized gas also became the starting concept to investigate plasma processing, and results have been interpreted mostly along this line.

In dealing with low-pressure plasma, one could argue that it is a partially ionized gas. However, use of the word *partially* has a limit based on at what extent it is partial. It is unreasonable to identify a system based on the minor component (e.g., less than number ratio 10^{-3}, which seems to be the case in most low-pressure plasma). It is certainly unreasonable to interpret the results based on the partially justifiable principle.

4.2 Luminous Gas Phase

It is clear that the term *plasma* was coined to identify the luminous gas phase, which is obviously visibly different from the ordinary gas phase. It was first recognized as "radiant (glowing) matter" and the concept of the fourth state of matter by W. Crookes 49 years before Langmuir coined the term *plasma*— that is, the recognition of a "luminous gas phase (glowing gas phase)" preceded the identification of a "plasma phase." If we focus on the nature of the gas phase, the luminous gas phase is based on what we conceive, but "plasma phase" as the ionized gas, used as the starting point of plasma processing, is based on the analytical data of matters that make the gas phase glow. Here lies the difficulty of describing the gas phase based on analysis of matters in the gas phase. Analysis generally depends on the method and the type of matter being analyzed. The analysis of the simplest organic molecule methane by mass spectrometer identifies a great number of species depending on the sensitivity of the equipment and ionization energy used (i.e., the gas phase, in general, cannot be described by analysis of a particular gas phase).

Even if it was granted that the gas phase created by glow discharge of argon in low pressure is ionized gas, ions of argon and electrons are not radiant (glowing) matters. The radiant matters, in this case, are excited species of argon atoms that emit photons. Therefore, if plasma state is viewed as an ionized gas, the interpretation of plasma processing, particularly plasma deposition, could become completely irrelevant to what happened by the process.

It is also important to recognize that the majority of studies on the fundamental physics of dielectric breakdown of gas phase in low pressure have been carried out with argon or other mono-atomic gases, which are ideally suited for verification of the ionization principle, simply because there is no complication due to dissociation and chemical reaction of the gas, which causes the loss of gaseous species.

The critically important difference between deposition plasmas and plasma of mono-atomic gases has not been recognized for a long time, mainly due to the perceived principle of ionized gas, which dictated the mind-set to interpret experimental results. As a matter of fact, the author had never looked at the plasma of monomers (gas used for plasma polymerization), with suspicion that it might be different from the plasma of argon in the same reactor, simply because of the perceived concept of plasma. Although the author had been convinced that the gas phase breakdown process of deposition gas had to be different from that of argon plasma, there was no hint what should be looked for and where one should look. The situation changed dramatically with the accidental discovery of "dissociation glow" [6,7] when the author was taking pictures of the luminous gas phase of a simple molecular gas of trimethylsilane to be used in his previous book [8].

Gases used in plasma polymerization are organic and inorganic molecular gases, of which the dissociation of molecules under the impact of electrons plays the key role, which is absent in mono-atomic gases. The loss of gas species due to the deposition leaves species that are not involved in plasma polymerization in the plasma phase, which makes the analysis of the gas phase irrelevant to plasma polymerization and could lead to misleading conclusions. With plasma polymerization of methane or trimethylsilane, the electron-impact dissociation of molecules yields large amounts of hydrogen, which become the main ionic species detectable in the plasma phase, but we are not polymerizing hydrogen to deposit.

The deposition of polymeric material from the vapor of monomers was recognized as plasma polymerization in early 1960 [9]. This term simply indicates that polymers are formed in the plasma of monomers. It has also been recognized that organic compounds that cannot be viewed as monomers of any type of polymerization can also be polymerized by plasma polymerization, which led to the reaction mechanism of repeating step growth polymerization (RSGP), which is explained in some detail in Chapter 7. The reaction mechanism is also applicable to the reaction mechanism of CVD in general. The only difference in the RSGP mechanism for the plasma polymerization and for CVD is the dissociation process that creates chemically reactive species dealt with in the RSGP growth mechanism (i.e., electron-impact dissociation in plasma polymerization, and thermal dissociation in CVD).

The recognition that the polymerization in plasma polymerization is not the polymerization in the conventional sense of polymerization, and also that plasma in plasma polymerization is not in the widely subscribed context of plasma as ionized gas, the process is described as luminous chemical vapor deposition (LCVD) [8]. The plasma polymerization and LCVD have been used synonymously, depending on where the emphasis is placed (e.g., plasma polymerization was used when the chemical growth mechanism is emphasized, and LCVD was used when the processing aspects are emphasized). LCVD seems to be the most straightforward expression of the process, considering all fundamental factors.

According to the widely subscribed mechanism of the gas phase breakdown [10,11], the primary electrons are created in the gas phase by the impact of naturally occurring ionizing radiation, such as of the cosmic ray. The electrons created in the gas phase follow the critical steps described below to cause the dielectric breakdown of the gas phase.

1. The electrons created in the gas phase are accelerated by the electric field in the gas phase and cause the ionization of the gas.
2. The ions are accelerated toward the cathode, and the bombardment of ions on the cathode surface causes the emission of the secondary electrons from the cathode surface.
3. The secondary electrons follow the same steps described for the primary electrons.

The details of these arguments are presented in Chapter 5.

The recent discovery of the cathode glow, which virtually touches the cathode surface (no dark space), in the deposition plasma of organic vapor [6–8,12,13], provided the key clue that the dissociation of molecules occurs prior to the ionization, which defies the above-mentioned classical mechanism of dielectric gas phase breakdown. The electrons that cause the dielectric breakdown of the gas phase are not the secondary electrons explained by the classical interpretation of the gas phase breakdown but are the primary (free) electrons pulled out of the surface state of the cathode metal [13,14]. Although the plasma phase of deposition gas can be described as a plasma phase, the ionization and ionized gas play minor roles. It is interesting to note that the pioneering group of plasma polymerization in early 1960 used "glow film" to describe their product. The description of plasma phase incidentally followed the historical pattern of the laboring effort to describe plasma phase. There is no objection to using the term *plasma* to describe the luminous gas phase in which plasma polymerization occurs, but the definition of *plasma*, in this case, is the luminous gas phase of which major component species are photo-emitting excited neutral species, and ionization or ions play a minor role, if any.

References

1. Crookes, W., "On radiant matter," lecture delivered to British Association for the Advancement of Science, Sheffield, UK, August 22, 1879; *Plasma Physics*.
2. Thompson, J. J., *Philosophical Magazine*, 44, 293, 1897.
3. Langmuir, I., "Oscillations in ionized gases," *Proceedings of the National Academy of Sciences of the USA*, 14, 628, 1928.
4. Tonks, L., "The birth of 'plasma,'" *American Journal of Physics*, 35, 857–858, 1967.
5. Mott-Smith, H. M. Letter to A. M. Bueche, April 20, 1967, Communications Operation, General Electric R&D Center, Schenectady, NY.

6. Yasuda, H. and Q. Yu, *Plasma Chemistry and Plasma Processing*, 24, 325–351, 2004.
7. Yu, Q., C. Huang, and H. Yasuda, *Journal of Polymer Science. Part A, Polymer Chemistry*, 42, 1042–1062, 2004.
8. Yasuda, H., *Luminous Chemical Vapor Deposition and Interface Engineering*, CRC Press, Boca Raton, FL, 2004.
9. Goodman, J., *Journal of Polymer Science*, 44, 551, 1960.
10. Chapman, B., *Glow Discharge Processes*, John Wiley and Sons, New York, 1980.
11. Brown, S. C., *Introduction to Electrical Discharges in Gases*, John Wiley and Sons, New York, 1966.
12. Yasuda, H., *Plasma Processes and Polymers*, 2, 293–304, 2005.
13. Yasuda, H., *Plasma Processes and Polymers*, 4, 347–359, 2007.
14. Yasuda, H., L. Ledernez, F. Olcaytug, and G. Urban, "Electron dynamics of low-pressure deposition plasma," *Pure and Applied Chemistry*, 80(9), 1883, 2008.
15. Rogoff, G. L., guest editorial, Special Issue on Applications of Partially Ionized Plasmas, G. L. Rogoff, Ed., *IEEE Transactions on Plasma Science*, 19, 989, 1991.

5

Dielectric Breakdown of Gas Phase

5.1 Significance of Dielectric Breakdown of Gas on Luminous Chemical Vapor Deposition

In order to perform the deposition of solid materials from the luminous gas phase in low pressure, which is generally described as low-pressure plasma, it is necessary first to create the luminous gas phase. Fundamental investigation of the dielectric breakdown of gas phase with molecular deposition gas is considerably more difficult than that of non-deposition gas, because the deposition severely hampers diagnostic measurement of the luminous gas phase, and the component species in the gas phase depends on the rate of deposition, which is dependent on the operation parameters. On the other hand, the breakdown occurs more or less spontaneously in many cases, and the gas phase breakdown is often taken for granted. Consequently, the understanding of gas phase breakdown of deposition gas, the monomer of low-temperature plasma polymerization, has been intuitively assumed to follow the same steps known for the dielectric breakdown of nondeposition gas, such as of argon in direct current (DC) discharge. It was recognized only recently that gas phase breakdown of molecules follows the reaction paths that are significantly different from those of mono-atomic gas (e.g., Ar) [1–6].

One factor that hampers thorough comprehension of the gas phase breakdown of deposition gas seems to be the dependence of the process on the modes of electrical discharge (e.g., the frequency of the electrical power source). The majority of investigations of low-pressure plasma polymerization are carried out by using a radio frequency (RF) power source, typically 13.5 MHz. However, most fundamental investigations of the dielectric breakdown of gas phase were investigated by DC discharge of mono-atomic gases (e.g., Ar), of which findings are not applicable to RF discharge because the motion of electrons in RF discharge is different from that in DC discharge. The dark cathode fall where electrons are accelerated by the electric field, the energy input plasma sheath in DC discharge, does not exist in RF discharge, because the polarity of the electric field changes so quickly that there is no space to accelerate electrons in linear motion as happens in DC discharge. However, there is a dark space in between the powered electrode and the main body of the luminous gas phase created by RF discharge. The dark space in RF discharge resembles the dark space in DC discharge and could be termed the electron accelerating dark space of RF discharge simply based on the resemblance, if one ignores or is not aware of the difference between what is happening in the dark space in RF discharge and in DC discharge.

The cathode fall dark space in DC discharge is the energy input sheath of the luminous gas phase created, and the main electron-impact reactions (with mono-atomic gases) occurring at the fringe of the dark space and the intensive glow. The negative glow near anode or dark space beyond negative glow is the energy-dissipating sheath of DC discharge. In contrast to this situation, the dark space in RF discharge, which is seen near the electrode surface, if used, is the sheath that separates the luminous gas phase created by RF discharge and the energy input surface. In the dark space in RF discharge, the acceleration of electrons does not occur, because the acceleration of electrons under a linear electric field does not occur with RF discharge. The electron-impact reactions with gases occur with oscillating electrons, which mainly occurs in the intensive glow adjacent to the sheath (dark space).

The modes of energy transfer to the gas phase are different in DC and in RF discharges. Consequently, the difference in the gas phase breakdown process with DC and that with RF discharge is vitally important in understanding the process of plasma polymerization. Furthermore, the distinction between nondeposition plasma and deposition plasma is of prime importance in understanding plasma polymerization. Those are the prerequisites in comprehending the magneto-luminous chemical vapor deposition (MLCVD) described in this book.

MLCVD utilizes a 15 kHz power source, which is customarily described as audio frequency (AF) discharge, under the influence of a magnetic field. Without a magnetic field, an AF discharge is an alternating DC discharge, which has two identical cathode fall dark spaces representing the cathodic cycle of the respective electrode. With a magnetic field superimposed to electrodes, the discharge mode changes significantly, and the cathode glow in the form of a toroidal glow develops, which is away from the cathode surface, somewhat resembling the situation with RF discharge—that is, the electrons do not accelerate from the cathode surface in a straightforward manner, but electrons circulating along the magnetic field (electrons in the magnetic field) cause, or participate, in the electron-impact reactions with gas molecules. The unique advantages of the MLCVD are attributable to the special mode of electron-impact reactions, which would become self-evident upon examination of the fundamental principle of the dielectric gas phase breakdown.

5.2 Breakdown of Nondeposition Gas and Deposition Gas

5.2.1 Dissociation Glow and Ionization Glow

The distinction between dielectric breakdown of nondeposition gas, such as Ar, and that of deposition gas, such as CH_4 and $HSi(CH_3)_3$, has not been well recognized. The breakdown process has been investigated mainly with Ar in DC discharge. It has been intuitively assumed that the principles found with Ar discharge would apply also to molecular gases in RF discharge. The conventional thought behind this approach that the gas phase created by glow discharge can be dealt with collectively as ionized gas has been challenged by the discovery of the dissociation glow with molecular gases in DC discharge that touches the cathode surface, in comparison to the glow that develops with mono-atomic gas, in which the glow is separated from the cathode surface by the cathode fall dark space.

The first glow develops with DC discharge of $HSi(CH_3)_3$, trimethylsilane (TMS), is the dissociation glow that attaches to the cathode surface (cathode glow)—that is, there is no cathode fall dark space in which electrons are accelerated to gain enough energy to ionize gas. The cathode glow means that electrons emanating from the cathode surface have enough energy to cause the dielectric breakdown of the TMS gas phase. Obviously, such a phenomenon cannot be explained by the Townsend–Paschen concept of gas phase breakdown.

Figure 5.1 depicts the experimental setup used to compare glows near the cathode surface with Ar and TMS. An aluminum plate cathode in the center is paired with two anodes equipped with a magnetic field behind the electrode plate. Two anodes are employed to make symmetrical glows on both side of the cathode, avoiding the stray glow that develops behind the cathode in an arrangement with two parallel electrodes. The effect of the magnetic field with an anode is discussed in a later section dealing with the influence of the magnetic field on the luminous gas phase (Chapter 6). In this section, let us focus on the location of glow near the cathode plate. Figure 5.2 shows the glow develop with Ar, which does not change with discharge time, indicating that what develops at the onset of glow remains as the steady-state glow.

Figure 5.3 schematically depicts electron-impact reactions in the dielectric breakdown of Ar. The primary electrons emanating from the cathode surface are shown as (⊖), and the electrons that came out of the Ar atom as a consequence of ionization are shown as (e) in order to distinguish the source of electrons. The presence of ⊖ in the ionization reaction is often ignored, probably because of the electroneutrality requirement in plasma. If ionization occurs solely by electron-impact ionization, the number of electrons is

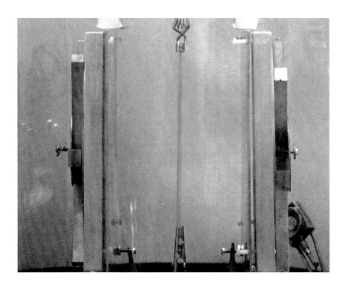

FIGURE 5.1
Electrode arrangement for investigation of glow near the cathode; aluminum plate plated in the middle of two anodes.

Ar DC Glow Discharge; Cathode Dark Space/Ionization Glow

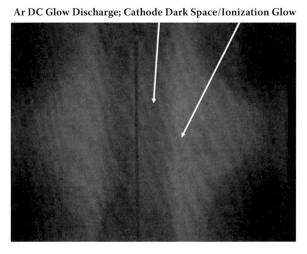

FIGURE 5.2 (*See color insert.*)
Glow develops with Ar DC discharge, with anode magnetrons.

greater than the number of ions, and the electron neutrality of the broken-down gas phase cannot be assumed. It is likely that recombination of electrons (e or $\ominus$) and ions constitutes the main path for the creation of excited species, as indicated in the figure, because the formation of excited species requires a narrow range of energy, but the probability of an electron colliding with an atom at the precise moment when the electron being accelerated

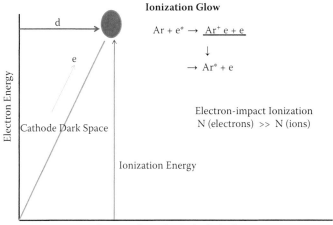

FIGURE 5.3
Glow location of Ar DC discharge.

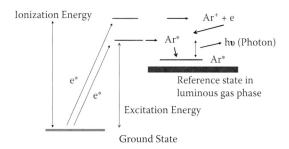

FIGURE 5.4
Energy level of species involved in luminous gas phase of Ar.

gained the precise energy necessary is very low. Figure 5.4 schematically depicts energy levels of various species with respect to ground state and the reference state in the luminous gas phase.

Figure 5.5 represents a pictorial view of the glow that develops at the onset of glow discharge (taken within 5 seconds of inception of glow) of TMS. Comparison of Figure 5.2 and Figure 5.5 clearly tells us these two processes are completely different. The main glow is the dissociation glow touching the cathode surface, and there is a faint ionization glow of atomic hydrogen that appears at nearly the identical location where the Ar glow appears. There is no dark cathode fall space between the primary glow and the cathode surface, which is clearly visible with Ar discharge. The absence of the cathode fall dark space for the dissociation glow indicates that electrons emanating from the cathode surface, as they enter the contacting gas phase of TMS, have

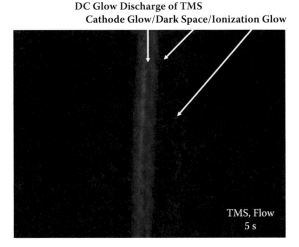

DC Glow Discharge of TMS
Cathode Glow/Dark Space/Ionization Glow

TMS, Flow
5 s

FIGURE 5.5 (*See color insert.*)
Glow develops with TMS DC glow discharge, with anode magnetrons.

sufficient energy to cause dissociation of TMS molecules. The main chemical reaction in the electron-impact dissociation of TMS is the homolytic splitting of C-H bonds, which releases hydrogen atoms.

There is clearly observable dark space between the dissociation glow and the ionization glow, which means that the primary electrons that did not collide with gas molecules as well as electrons that delivered energy to create dissociation glow are accelerated in this dark space, as in the case of Ar discharge, but in the case of TMS, the main ionizable species are the products of the electron-impact dissociation reactions, which took place in the dissociation glow. Thus, the electron-impact ionization of the dissociation products, not the original gas, occurs. Figure 5.6 schematically depicts the

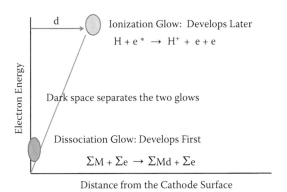

d — Ionization Glow: Develops Later

$$H + e^* \rightarrow H^+ + e + e$$

Dark space separates the two glows

Dissociation Glow: Develops First

$$\Sigma M + \Sigma e \rightarrow \Sigma Md + \Sigma e$$

Electron Energy

Distance from the Cathode Surface

FIGURE 5.6
Two glows that develop with TMS DC discharge.

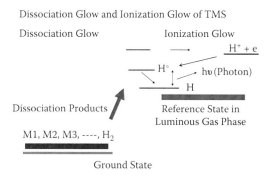

FIGURE 5.7
Energy levels of species in dissociation glow and ionization glow of TMS DC discharge.

developments of dissociation glow and ionization glow in DC discharge of TMS. Figure 5.7 schematically shows the energy levels of species involved in dissociation glow and ionization glow of the TMS DC discharge.

The glows that develop with TMS (deposition gas) consist of the primary glow (dissociation glow) and the secondary glow (ionization glow), and the intensity of the secondary glow depends on the increase of the concentration of the dissociation products from the primary glow. Consequently, the intensity of the secondary glow is dependent on the reaction time. Figure 5.8 depicts pictorially the change of glows with reaction time for closed-system and flow-system experiments. In the closed system, the dissociation glow disappears, and only the ionization glow remains in 180 s, while in the flow

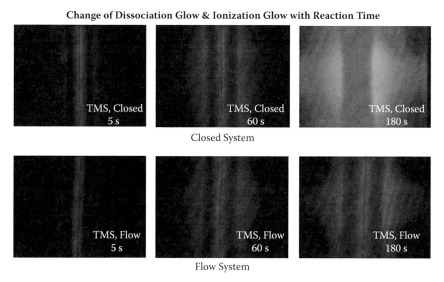

FIGURE 5.8 (*See color insert.*)
Change of dissociation glow and ionization glow of TMS with reaction time.

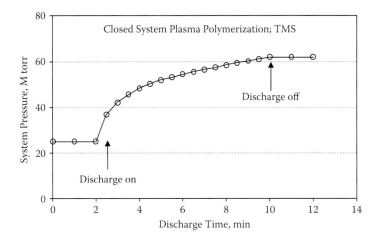

FIGURE 5.9
Change of pressure with reaction time of TMS closed-system DC discharge.

system, both the dissociation glow and the ionization glow reach the dynamic steady state within a relatively short time. Figure 5.9 shows the change of system pressure with reaction time of TMS DC discharge in a closed system. DC discharge of Ar does not change system pressure regardless of whether in a closed system or flow system.

The system pressure in a closed-system discharge of TMS increases in spite of the fact that the majority of mass leaves the gas phase by the deposition of plasma polymers of TMS. This occurs because the overall gas increases due to the dissociation of TMS molecules, and pressure is determined by the number of gaseous species regardless of their mass. Figure 5.10 depicts the change of

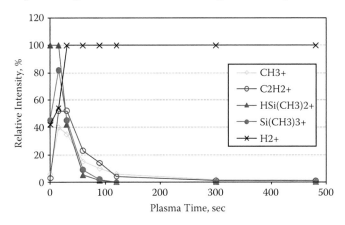

FIGURE 5.10
Change of species in the luminous gas phase of TMS with reaction time in a closed system.

gaseous species (photoemission spectroscopy) with reaction time. The major detectable species of TMS dissociation disappear within 120 s (deposition completes), and the remaining gas in the system becomes H_2.

Figure 5.11 depicts main photo-emitting species in dissociation glow and ionization glow of TMS in the dynamical steady-state luminous gas phase. This figure clearly shows that the dissociation of TMS occurs in the dissociation glow, and the ionization glow is due to the ionization of H atoms, which are major products of the dissociation reactions that remain in the closed gas phase. Thus, it is quite obvious why Ar discharge does not yield deposition of material, and TMS discharge yields deposition of plasma polymer of TMS (i.e., the dissociation of gas is necessary to cause the deposition of materials). Conversely, detailed investigation of the gas phase breakdown of Ar does not provide meaningful and necessary information for understanding deposition plasma.

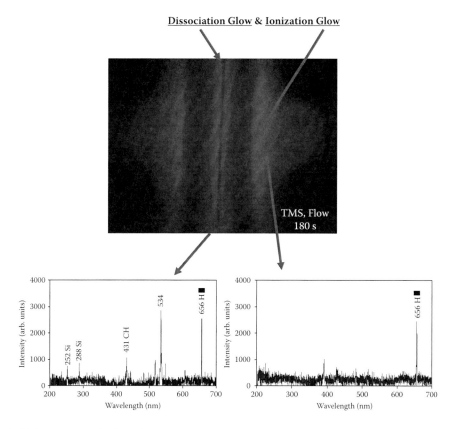

FIGURE 5.11 (*See color insert.*)
Major photon-emitting species in dissociation glow and ionization glow of TMS.

5.2.2 Interrelationship of Dissociation Glow and Ionization Glow and Influence of Power Source

The dielectric breakdown of deposition gases, organic molecules, and the gas involved in the ionization glow are products of the dissociation of original gas, which occurs in the dissociation glow, as shown in Figure 5.11. However, when a molecular gas (e.g., TMS) is mixed with oxygen, glows that develop at the onset of gas phase breakdown are the dissociation glow of TMS and the ionization glow of O_2, as shown in Figure 5.12 (left) with the blue dissociation glow of TMS and the gold ionization glow of O_2. However, reactive species in these different glows interact, and the glow changes within a few seconds. The respective colors for the dissociation glow and the ionization glow changes are as shown in Figure 5.12 (right). A few seconds after the onset of glow discharge, the original ionization glow of O_2 changes to the ionization glow of H_2, which is the dissociation product of TMS, because oxygen reacts with Si to form stable Si-O containing moieties.

Knowing that TMS dissociates by electron-impact dissociation, it is important to examine how the dissociation appears with 15 kHz AF discharge and 23.8 MHz RF discharge. Figure 5.13a depicts the difference of dissociation glow of TMS. Because 15 kHz AF discharge is essentially DC discharge

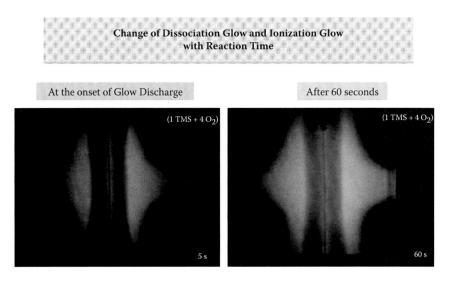

FIGURE 5.12 *(See color insert.)*
Change of dissociation and ionization glows due to the interaction of reactive species in dissociation and ionization glow. The ionization glow of O_2 changes to that of H_2, which is one of the dissociation products of TMS.

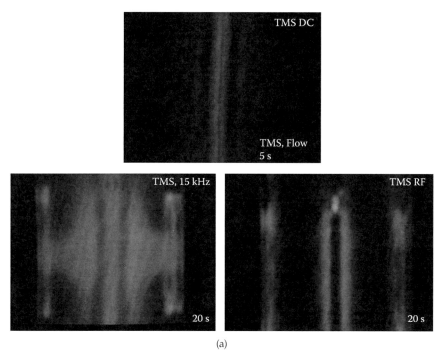

(a)

FIGURE 5.13 (*See color insert.*)
(a) Effect of frequency of power source on the dissociation glow of TMS.

with alternating polarity, the dissociation glow that appears on the center electrode without a magnetic field is identical to the DC discharge shown on the upper row. The most striking difference is seen with RF discharge (i.e., the center electrode is completely in dark space, and there is appreciable distance between the electrode surface and the edge of the intense glow). As mentioned in the preceding section, the space surrounding the center electrode is not the same dark space that separates the cathode surface and the intensive glow (ionization glow) of Ar DC discharge, which is the cathode fall dark space where electrons are being accelerated in DC discharge. In RF discharge, electron-impact dissociation of molecules takes place between oscillating electrons and gas molecules, and the dark space does not have the same function as the cathode fall space in DC discharge. Often, this sheath is misinterpreted as the sheath of plasma in the context of DC discharge of Ar. Figure 5.13b shows a similar dependence on the power source with $C_2H_2F_2$. All aspects described for discharge of TMS are also seen with $C_2H_2F_2$, except the glow expands much wider than what is observed with TMS, which is discussed in Chapter 6.

$C_2H_2F_2$

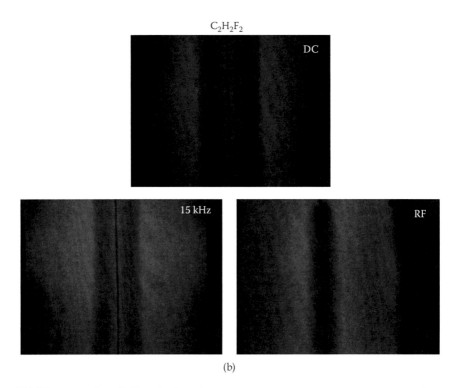

(b)

FIGURE 5.13 (continued). (*See color insert.*)
(b) Effect of frequency of power source on the dissociation glow of $C_2H_2F_2$.

5.3 Source of Electrons for Electron Avalanche to Cause Gas Phase Breakdown

It is vitally important to understand where and how electrons that cause electron-impact reactions, including dissociation, excitation, and ionization, originate. There are at least two schools of thought for the origin of electrons.

5.3.1 Secondary Electrons Emission by the Bombardment of Accelerated Ions Caused by the Ionization of Gas in the Gas Phase; Townsend's Gas Phase Ionization Hypothesis

This concept is based on the ionization of gas (e.g., Ar) occurring by impact of naturally occurring ionizing radiation [7,8]. The gas ions created are accelerated toward the cathode and cause emission of the secondary electrons. The electrons created by the ionization in the gas phase are considered as the primary electrons. The bombardment of the accelerated ions on the cathode surface is considered to cause the emission of the secondary electrons, which will be accelerated in the electric field to gain enough energy to ionize the gas to create and sustain the luminous gas phase.

In this scheme, the essential gas phase breakdown occurs as a consequence of bombardment of positive ions on the cathode metal surface and the consequential emission of the secondary electron. It is considered that the first gas phase ionization by an undefined process such as irradiation of a cosmic ray is necessary to create ions in the gas phase. However, the theory is vague on the role of electric field applied in the effort to cause the gas phase breakdown and on which step the electron avalanche occurs to create the luminous gas phase. This approach mainly explains the characteristics of the broken-down gas phase in terms of parameters involved.

5.3.2 Free Electrons Emission from the Surface State of Cathode Metal by the Applied Electric Field; Primary Electron Emission Principle by Yasuda

This concept is based on the interfacial electron transfer of free electrons. The secondary emission scheme explained above does not refer to the nature of electrons emitted by the bombardment of energetic ions. Dealing with electrons, it is important to note that the perception of electrons largely depends on the discipline in which scientists work. Chemists are mainly concerned with valence-level electrons, physicists mostly core-level electrons, biochemists unpaired electrons in the form of free radicals, and electronics engineers Fermi-level electrons. However, there are abundant quantities of free electrons that are not bound in any form of restriction thus mentioned; hence, not much attention has been paid to each specific topic of discipline. Free electrons seem to exist in various materials. If those free electrons in the surface state of cathode metal could be pulled out by the electric field applied between the cathode and the anode, the primary electrons could be emitted from the cathode surface and could be accelerated in the applied electric field that caused the emission. In such a case, the same driving force, applied electric field, explains both electron emission and acceleration of emitted electrons.

This concept stemmed from the observation of the interfacial electron transfer that occurs when contact is made between two surfaces, which occurs without an external driving force [9,10]. An extension of this observation leads to the question, "can free electrons in the surface state of cathode metal be pulled out by the applied electric field?" The answer is "yes," which is explained in the following sections dealing with the dielectric gas phase breakdown, the key mechanism of the primary electron emission.

The accelerated electrons could cause the electron-impact reactions irrespective of the mechanisms of electron emission from the cathode surface. The only and the major differences are the role and sequence of the formation of ions. In this concept, ions are created by the electron-impact reactions (i.e., the primary electron emission causes the formation of ions). According to Townsend's hypothesis, the ions formed in the gas phase (by cosmic ray) cause the emission of the secondary electrons. This discrepancy can be clarified only by examining the applicability of the concept to the dielectric breakdown process of deposition gases, which has not been touched by conventional analysis of the phenomena, simply because nearly all investigations were performed with mono-atomic gas (e.g., argon).

5.4 Interfacial Electron Transfer

5.4.1 Interfacial Electron Transfer in the Static Charge Creation in the Contact Electrification

The static charge created by making contact between two dissimilar materials and separating the contact occurs due to the interfacial electron transfer, which could be explained by the following schematic presentations. Figure 5.14 depicts the movement of free electrons in the surface state of the two materials when contact is made. Figure 5.15 depicts the overall change of electrons in the surface state of two materials on the sequence of contact and separation. Each surface state has a certain number of electrons based on the previous equilibration with the contacting medium and characteristic overall electronegativity of the surface, which is not a specific number attributable to a specific material but depends on the previous history of a material (as depicted in the top diagram in Figure 5.15).

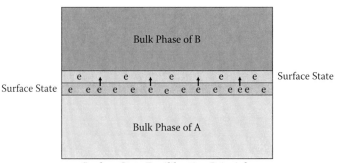

FIGURE 5.14
Interfacial electron transfer.

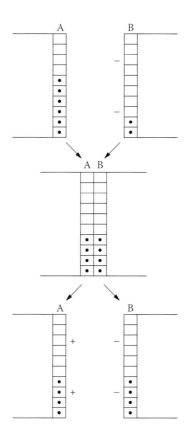

FIGURE 5.15
Static charge buildup by contact and separation.

When a contact is established, the two surface states follow the equilibration of surface states with respect to the number of electrons in the surface state (as shown in the middle diagram of Figure 5.15), with the equal number of electrons in both surface states. In Figure 5.15, it is simplified using the number of electrons, but in real cases, the equilibration depends on the degree of saturation of each surface state, which depends on the history of each surface prior to making contact. Then two surfaces are abruptly separated after a certain duration of contact, and two surface states retain the newly established equilibrium number of electrons. The surface that gained electrons charges negatively, and the surface that lost electrons charges positively.

The interfacial electron transfer explained above can be quantitatively examined by completing an electric circuit from the contacting surface to the back surface of the test surface that contacts with a reference surface. The transfer of electrons causes the current, which is designated as the contact current. The transfer of electrons on the separation process is designated as the separation current. Because of the short-circuiting, the charge buildup

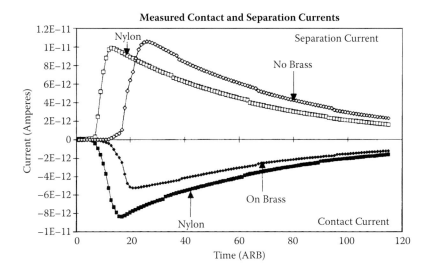

FIGURE 5.16
Contact current and separation current.

on the surfaces does not occur in this experiment. Figure 5.16 depicts contact current and separation current observed with two different surfaces contacting the reference surface. The time scales of plots are shifted in order to identify the two cases separately. The figures clearly show that the interfacial transfer of electrons occurs. The contact current and the separation current are nearly identical, except the sign is opposite.

The surface used for this experiment was prepared by depositing approximately 30 nm of plasma polymer of methane on a brass plate and on a nylon film, and an uncoated brass plate was used as the reference surface. This experiment also shows the following important characteristics of the surface state. The tribo-electric characteristics of 30 nm thick plasma polymer deposited on brass plate and on nylon film are identical, which means that plasma polymer can be used as implanted or surface-grafted surface state, independent of the substrate; the effective thickness of the surface state could be viewed as roughly 30 nm; and the effective plasma polymerization coating could be obtained with the deposition of 30 nm or less, which agrees with the effect of thickness described in a later section.

5.4.2 Correlation between Number of Free Electrons in the Surface State of the Cathode Metal and the Breakdown Current

If the free electrons in the surface state of metal used as the cathode can be pulled out by the applied electric field that develops at the onset of glow discharge, the current should be proportional to the number of free electrons in the surface state of the metal used as the cathode. It was found that the

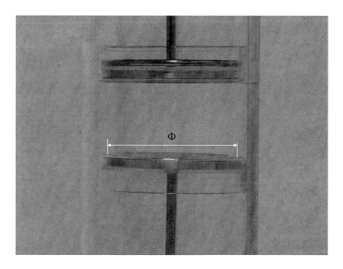

FIGURE 5.17
DC discharge reactor used for nondeposition plasma.

breakdown voltage depends on the metal used as the cathode, but its extent is marginal, and the breakdown current (the current observed at the onset of gas phase breakdown) depends significantly on the kind of metal [6]. This is the first indication that the breakdown current is far more important than the breakdown voltage, which has been dealt with as the sole parameter in investigation of the dielectric breakdown of the gas phase in the conventional approach.

Figure 5.17 depicts a simple reactor with two electrodes, which was used to examine the effect of cathode metal on the breakdown process. Each electrode surface is covered with a Pyrex glass petri dish, on which a circular hole with a defined diameter is made on the bottom surface. By changing the size of the hole in the petri dish, the size of the cathode surface area was changed without changing the size of metal electrodes. The petri dish also covers the edge of the metal electrode and eliminates the edge effect of the electrode (i.e., if the edge is exposed, the electric field lines concentrate at the edge and cause the higher intensity of glow at the edge).

Figure 5.18 depicts the correlation between the estimated number of free electrons within the surface state of the cathode and the current observed at the onset of glow discharge [6]. The figure seems to strongly support the concept that free electrons are pulled out by the electric field applied. Because the dependence of the breakdown voltage on the types of cathode metal is marginal, the energy transferred to the luminous gas phase is proportional to the breakdown current, which is dependent on the number of free electrons in the surface state of the cathode. Figure 5.18 clearly shows the linear dependence of the current at the onset of gas phase breakdown on the estimated number of free electrons in the surface state of metal.

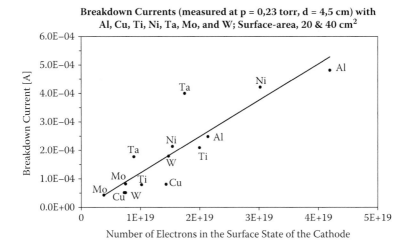

FIGURE 5.18
The dependence of breakdown current on the number of electrons available in the surface state of metals used as the cathode.

In the estimation of the number of free electrons in the surface state of cathode metal, it is assumed that the number of free electrons that can be in the surface state is proportional to the number of electrons on the outermost orbit of the metal used as the cathode. Two methods of estimation were used, which yield a similar trend, and the thickness of the surface state, 30 nm, was used for the *x*-axis of the plots. Table 5.1 tabulates the number of electrons

TABLE 5.1

Calculated Density of Electrons in the Cathode
by Different Models

Density of Electrons in the Cathode n_e (in 1/m³)	From the Metallic Radius	From the Drude–Sommerfeld Model
Aluminum	2.45×10^{29}	1.81×10^{29}
Copper	1.14×10^{29}	8.49×10^{28}
Molybdenum	8.70×10^{28}	6.40×10^{28}
Tantalum	1.50×10^{29}	1.10×10^{29}
Titanium	1.53×10^{29}	1.13×10^{29}
Tungsten	1.70×10^{29}	1.26×10^{29}

Sources: The metallic radii are taken from Marder [M. P. Marder, *Condensed Matter Physics*, Wiley-Interscience, NY, 2000]. The Drude model calculation was taken from Ashcroft et al. [N. W. Ashcroft and N. D. Mermin, *Solid State Physics*, Saunders College, Philadelphia, PA, 1976].

on the outermost orbitals of seven metals used. The fact that the breakdown current is dependent on the number of electrons clustering in the surface state of metal used as the cathode is strong evidence in support of the electric field emission of the primary electrons for the gas phase breakdown.

5.5 Experimental Examination of Gas Phase Breakdown

5.5.1 Parameters That Influence Gas Phase Breakdown

The surface-state electrons, unbound (free) electrons in the surface state of the metal, interact with the contacting gas; however, it is to a generally minuscule extent without applied electrical field across the gas phase. When a DC voltage is applied between two electrodes separated by a gas phase, a minute current can be measured, which is termed the *dark current*, because the current flows through without causing the dielectric breakdown of the gas phase. When the applied voltage is raised slowly, the dark current remains at a constant value until dielectric gas phase breakdown occurs. Figure 5.19 schematically depicts the change of voltage and current at the onset of the breakdown of the gas phase.

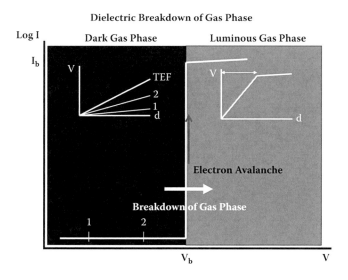

FIGURE 5.19 (*See color insert.*)
The dielectric gas phase breakdown in terms of applied voltage and observed electric current: *V–d* plots (insert) show the change of electric field, which is the slope; *V/d*, where *d* is the separation distance of the cathode and anode.

The dielectric breakdown of the low-pressure gas domain involves the change of electric field as shown schematically in Figure 5.19; the increase of current at the onset of an electron avalanche is in the order of 10^6 to 10^7. The electric field, V/d, in the dark gas phase is an imaginary number, until the threshold electric field (TEF) is reached and electron avalanche occurs; at this point, electric current flow through the gas phase (i.e., the gas phase becomes an electron-transporting medium). At TEF, the transformation of the dark gas phase (left side) to the luminous gas phase (right side) occurs, which accompanies the change of the electric field profile associated with the breakdown process. The electric field shifts to V/d_L, with d_L indicated by an arrow on the right side of the diagram. The actual electric field that accelerates electrons emanating from the cathode surface is not that given by (V_b/d) but the slope (V/d_L).

When the electric field reaches a threshold value for the dielectric gas phase breakdown, the surface-state electrons are pulled out of the surface, which is the emission of the primary electrons by the electric field. The emission of the primary electrons occurs in the mode of electron avalanche, which is not emission of the secondary electrons caused by the bombardment of ions. Those experimental data shown above support the concept.

Since the interfacial transfer of electrons occurs without external driving force, as evidenced in the tribo-electric electron transfer, the free electrons clustering in the surface state of metal can be easily pulled out by applying electric field in the mode of electron avalanche.

The applied voltage that causes the electron avalanche is recognized as the breakdown voltage, V_b, and the (sustainable) current at the onset of gas phase breakdown is recognized as the breakdown current, I_b, in this book. These are practical and meaningful terms that describe the gas phase breakdown of deposition gases. The same terms are used in a much more restricted manner in handling the details of the physics of the breakdown process; however, in a practical sense, glow discharge of deposition gases can be initiated more or less spontaneously when applied voltage reaches the threshold value. More important is the consequence to the gas—that is, what happens to the gas? This is the key issue that separates molecular deposition gas from mono-atomic nondeposition gas.

5.5.2 Breakdown Voltage According to Townsend–Paschen Hypothesis

5.5.2.1 *Townsend Theory of Dielectric Gas Phase Breakdown*

The prevailing theories or experimental approaches on the dielectric breakdown of the gas phase in low pressure are modifications or extensions of the original concept of Townsend [7,8], which generally explains the broken-down gas phase but does not address the mechanism for the transformation from the dark, dielectric gas phase to the luminous, electrically conducting gas phase. The gas phase breakdown can be highlighted as follows. (Pay special attention to those aspects expressed in italics.)

- *Some electrons (primary electrons) are present in the gas phase.*
- Accelerated by the applied electric field, *the primary electrons* gain energy, and after a given path, they *can reach the ionization energy of the neutral gas.*
- Collision with the gas then *creates another electron and ion.* Those accelerate along the field line, with the electrons moving toward the anode and further colliding, and the ions moving toward the cathode.
- The probability of an electron colliding with and ionizing a neutral is decribed by A (first Townsend coefficient).
- The ions hitting the cathode with sufficient energy (twice the work function) release a "secondary" electron. This event is called *secondary electron emission*, and its probability of happening is described by B (second Townsend coefficient).
- Those two coefficients depend on *the electric field (applied voltage divided by the interelectrode distance)* and the pressure, for a given gas in a given reactor.

Figure 5.20 shows the general formula for this correlation. Based on this correlation, it has been generally considered that the breakdown voltage, V_b, of the dielectric breakdown of gas phase in DC discharge is dependent solely on the product of gas pressure, p, and the separation distance of the cathode and the anode, d—that is, $V_b = f(p * d)$ as Paschen's law [7,8,11–13]. However, many researchers, in recent years, found that V_b does not depend solely on the value of $p * d$ [14–21]—that is, Paschen's equation does not hold, but no alternative mechanism, beyond modification of parameters in the equation, for the dielectric breakdown of gas phase has been presented. Auday and coworkers found that multiple Paschen curves (plots of V_b against $p * d$) were

Paschen/Townsend Theory

- 1915: Townsend developed a theory based on ionization collision probability. From the theory an equation was later derived:

$$V_b = \frac{A^* \, p^* \, d}{B + \ln(p * d)}$$

FIGURE 5.20
Equation that correlated the breakdown voltage, V_b, to the Paschen parameter, (p^*d), where p is the system pressure, and d is the separation distance of the cathode and the anode.

obtained by changing the separation distance, d, while all other experimental parameters were unchanged, $V_b \neq f(p * d)$ [16]. Hassouba and coworkers showed that breakdown voltage of Ar and He was dependent on the type of metal used as the cathode [20].

It is worth noting that in most studies of breakdown voltage by means of Paschen plots, the separation distance d was not changed, and the values of breakdown voltages were plotted against $p * d$ by using a constant value of d. This trend is probably due to the fact that d in many experimental setups is a fixed reactor parameter, which cannot be dealt with as an experimental variable. It is also noteworthy that the original study by Paschen was for the streaking breakdown of gas separated by two balls (radius up to 1 cm), and the separation distance up to 1 cm. Most breakdown studies with parallel plate electrodes which appeared in the literature were carried out within a range of d less than 1 cm. Hence, it is obvious that the conventional view of the gas phase breakdown based on the Paschen–Townsend gas phase ionization theory does not adequately represent the transformation of the dielectric gas phase in between two parallel electrodes as a whole to the laterally uniform (nonstreaking) electrically conducting luminous gas phase, such as those dealt with in plasma polymerization reactors.

5.5.2.2 Flaws in Paschen–Townsend Treatment of Gas Phase Breakdown

The gas phase equation tells that the gas pressure is determined by the amount of gas but not by the weight of gas. In electron-impact reactions, on the other hand, the energy carried by electrons is transferred to the mass of gas (i.e., the influence of electron impact is dependent on the mass of gas). The important issue is the consequence of the electron impact to the gas. The deposition of materials from deposition plasma also depends on the mass of the depositing entities, the products of the electron-impact reaction with gas.

It is important to recognize that nearly all breakdown studies were carried out with simple atomic gases (e.g., He, Ar, etc.), and practically no study has been done with molecular gases used for plasma polymerization. This is a natural occurrence, because the deposition of materials from the luminous gas phase created by electrical discharge did not become a phenomenon of interest until the mid 20th century, although it had been recognized nearly 100 years before, but as a nuisance. Because no depositable species is created from Ar, the information obtained by the gas phase breakdown of Ar might have no relevance to that of deposition gases. In other words, the knowledge gained from the gas phase breakdown of Ar or He should not be extended to the gas phase breakdown of molecular gases. Only the general principle could be used to investigate the gas phase breakdown of molecular gases. However, the general principle derived from the Paschen–Townsend hypothesis cannot be used for this purpose because of the flaws described below.

Classical treatments of the dielectric gas phase breakdown have been carried out without describing the gas phase system in which the dielectric breakdown of the gas phase occurs. If the interelectrode distance d is changed, the volume between two electrodes also changes, and the pressure dependences of breakdown voltages in different gas systems are compared in the approach based on the Paschen plot. It is natural to find multiple Paschen curves by changing the separation distance d simply because phenomena in many different systems are compared, and curves representing each system are obtained.

5.5.2.3 Important Factors Not Considered in Paschen–Townsend Interpretation of the Gas Phase Breakdown

1. The electric field profile changes upon the breakdown of the gas phase, as depicted in Figure 5.19.
2. The plasma sheath, in which electrons and ions are accelerated, does not exist in the dark gas phase.
3. Recombination of electron and ion is a spontaneous reaction. How could ions keep bombarding the cathode surface against the flux of the secondary electrons?
4. Transformation from the dark gas phase to the luminous gas phase cannot be investigated by using the parameters of the luminous gas phase according to the fundamental principle of reaction kinetics (i.e., parameters of the dark gas phase control the transformation).

5.5.2.4 Breakdown Voltage According to the Dark Gas Phase Parameters

According to the gas law, the gas system can be defined by the following factors:

1. The gas phase can be defined by the equation $pV = nRT$, where p is pressure in Pa, V is volume in liters, n is the amount of gas, R is the gas constant, and T is the temperature in Kelvin.
2. One mole of gas, which consists of Avogadro's number, 6.02×10^{23} gas, occupies 22.4 L at standard pressure (1 atmosphere = 76 torr = 1.013 $\times 10^5$ Pa) and standard temperature ($0°C = 283$ K). When n is given by the number of moles, the molar gas constant $R = 8.31$ J/mol•K.
3. Mass of gas is given by M = (number of mole) × (molecular or atomic weight).

If one defines the electric field that causes the dielectric breakdown of the gas phase as the threshold electric field (TEF), which can be given by V_b/M_v, where M_v is the total mass in the interelectrode volume, the gas phase breakdown could be shown as the plot of the TEF against M_v:

(Threshold electric field in dark gas phase) = F (mass in the dark gas phase)

Accordingly,

$$V_b/(p * d * S * M/RT) = F(p * d * S * M/RT)$$

This equation describes the principle for the dielectric breakdown of the gas phase. How this relationship holds can be examined using the data of Auday and coworkers and also the data of Hassouba and coworkers.

Figure 5.21 shows Paschen plots of data presented by Auday and coworkers. Figure 5.22 presents TEF plots of the same data. Figures 5.23 and 5.24 are for the data from Hassouba and coworkers. These plots show multiple Paschen curves rather than a single curve that represents Paschen's law.

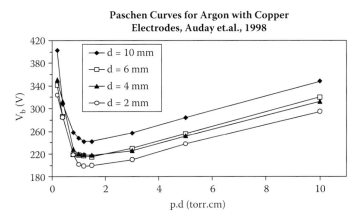

FIGURE 5.21
Paschen curves obtained at different values of *d*.

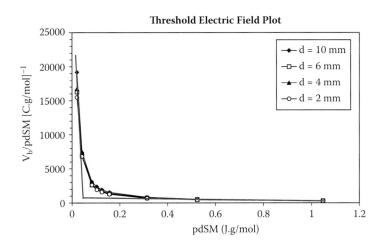

FIGURE 5.22
Threshold electric field plots of data shown in Figure 5.21.

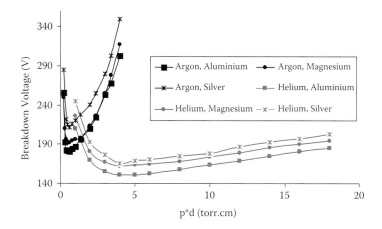

FIGURE 5.23
Paschen curves for Ar and He obtained with three different metals as cathodes. (Hassouba et al. 2002)

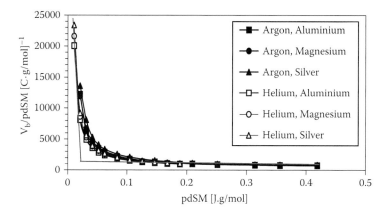

FIGURE 5.24
Threshold electric field plots of data shown in Figure 5.23.

Figure 5.25 shows multiple Paschen curves (using p as the x-axis parameter, at a fixed d) obtained when parameters that define the reactor gas phase are changed (i.e., size of electrode, type of gas, metal used as electrode, and gas pressure while maintaining a constant value of d). These Paschen curves clearly show that Paschen curves change with any parameter change that describes the gas phase, which proves the invalidity of Paschen plots for evaluation of the breakdown process in plasma polymerization reactors. Figure 5.26 presents the TEF plots of data shown in Figure 5.25. The TEF plots show more universal features of the dielectric gas phase breakdown phenomena. There are two domains, at higher pressure and at lower pressure,

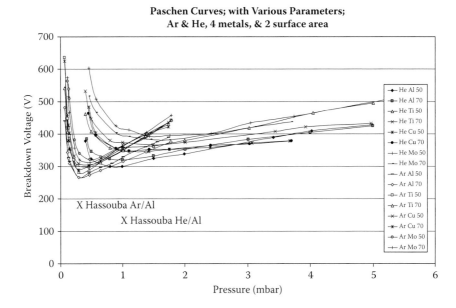

FIGURE 5.25 (*See color insert.*)
Paschen plots at a fixed *d*, two gases, four metals, and two surface areas.

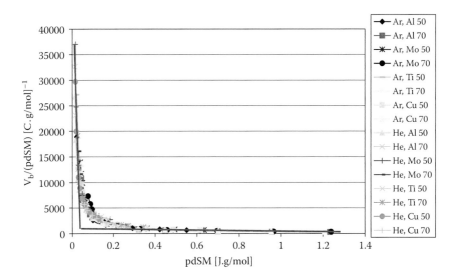

FIGURE 5.26 (*See color insert.*)
Threshold electric field plots of data shown in Figure 5.25.

and a transitional domain between those two domains, which justify the phase diagram of gas phase breakdown described in Section 5.8.

In most practical cases, the only variable parameter of operation is pressure. Therefore, in the following sections, breakdown voltage is plotted against pressure in order to simplify comprehension of the influence of major factors not previously considered.

5.5.3 Breakdown Process Investigated by Breakdown Voltage, Breakdown Current, and Derived Parameters

According to the classical interpretation, based on Paschen curve and Townsend's hypothesis of gas phase ionization, the gas phase breakdown has been treated as the ionization process, and accordingly, only the breakdown voltage has been examined to describe the dielectric gas phase breakdown. The dielectric breakdown of the gas phase could be viewed, alternatively, as the electron-impact excitation of gas to create excited species (inclusive of ions) that function as the energy-transporting medium. The photo-emitting neutral species could serve as more efficient energy-transporting medium than ions. The energies necessary for forming and maintaining photon-emitting excited species require significantly less energy than those for ions. The repeated reexcitation of species, which lost energy by emitting photons, by low-energy electrons, which is described in Chapter 6, enhances the energy-transferring aspect of the luminous gas phase.

The concept of the luminous gas phase as an energy-transporting medium could be explained by an analogy of how the energy of a tsunami is transferred by water; water acts as the energy-transferring medium, while water is not being transported. The efficiency of the energy-transporting medium could be judged by the conductivity of the broken-down gas phase, which requires the values of breakdown voltage and breakdown current, which are measurable parameters in DC discharge.

Any electron-impact reaction beyond the ionization of a mono-atomic gas such as Ar depends on the type, or nature, of gas. Hence, the investigation of the dielectric breakdown of gases as a function of the type of gas is necessary, for example, mono-atomic gas such as He and Ar, simple molecular gas such as N_2 and O_2, and organic molecules such as CH_4 and C_2H_2. The detailed investigation of the physics of the dielectric gas phase breakdown was carried out with argon nearly exclusively. It has been intuitively assumed that the principle found with argon could be applied to any other gases. Because argon, a mono-atomic gas, has no capability to participate or cause a chemical reaction, it is easy to diagnose the change of gas phase. The ionization is the major factor in creating the conducting luminous gas phase of argon, because it is likely that the photo-emitting excited neutral species are formed by the recombination of an electron and an ion.

However, with molecular gas (e.g., O_2, N_2, CH_4, etc.), the dissociation of molecules in the gas phase, which is absent in the case of mono-atomic gases, becomes the predominantly important factor in the breakdown of the gas phase. The major constituent species in the broken-down gas phase are excited neutrals; neutral species outnumber ions 10^6 [22]. Furthermore, the key operational parameter of DC discharge of deposition gases is the current density [23]. Therefore, the breakdown process should be investigated with both breakdown voltage and breakdown current, which is possible with DC discharge or lower-frequency alternating power–driven discharge.

The majority of plasma polymerization, particularly laboratory experiments, is carried out with RF discharge, in which voltage and current are not measured and the only power input parameter available is wattage, which might have been the factor to justify using the principle obtained by mono-atomic gas in DC discharge. Therefore, it is necessary to examine DC discharge with respect to breakdown voltage, V_b, breakdown current, I_b, and breakdown wattage, W_b, in order to gain knowledge that is applicable to those different types of discharges. The deposition of materials from the luminous gas phase severely hampers the application of investigations that could be easily carried out with Ar to the gas phase breakdown of molecular deposition gas. Consequently, it is necessary to rely upon the comparative studies of the breakdown phenomena with various groups of gases.

Some factors that are against the gas phase ionization (the secondary electron emission) but in favor of the emission of primary electrons are as follows:

1. The electric field that accelerates electrons to gain enough energy to ionize gas is not high enough to accelerate ions to cause the secondary electron emission, which is obvious when one considers the difference in size and mass of Ar and electrons.

2. Ions cannot bombard the cathode surface against a high flux of electrons under the conditions of creating glow discharge of a gas in low pressure. The recombination of a positively charged ion and an electron is a spontaneous reaction prompted by a Coulombic attractive force.

3. The secondary electron emission concept is vague on when and how the dielectric breakdown of the gas phase occurs but mainly describes the characteristics of the broken-down gas phase, while the primary electron concept is clear that the electron avalanche from the cathode surface is the triggering step of gas phase breakdown.

After all, the secondary electron emission concept has not been tested with deposition gases. It is necessary to comprehend the fundamentals of the interfacial electron transfer that points out the importance of electron affinity, or electronegativity, of atoms in order to test the primary electron emission concept.

5.6 Factors That Control Transformation of Gas Phase

5.6.1 Parameters of Reaction Kinetics

In order to elucidate the mechanisms for the transformation from the dark insulating gas phase to the luminous conducting gas phase, it is mandatory to identify the driving force of the process. The reaction kinetics principle of the transformation of dark dielectric gas phase to luminous electrically conductive gas phase is schematically shown in Figure 5.27. According to the reaction kinetics principle, the transformation is governed by parameters in the dark gas phase. Conversely, the reverse process (extinction of plasma) is dependent on parameters of the broken-down gas phase. This principle indicates that the Paschen–Townsend approach cannot explain the gas phase breakdown process. The detailed analysis of the luminous gas phase does not lead to the elucidation of the mechanisms of transformation of gas phase that created the luminous gas phase.

The Townsend equation can serve as a curve-fitting equation bound to a specific set of conditions of gas phase breakdown but cannot predict the phenomena in a generic sense. The detailed analysis to obtain parameters A and B, shown in Figure 5.20, does not lead to a prediction for what would be the breakdown voltage curve when new experimental conditions are employed. It is important to recognize that the electric field calculated by V/d, where d is the separation distance of cathode and anode, is not the electric field that pulls out electrons from the cathode surface, as depicted in Figure 5.19. The electron avalanche occurs concurrently with the change of V/d to V/d_L, where d_L is the width of the cathode fall dark space (i.e., V/d_L is the electric field that creates the luminous gas

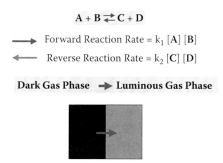

$$A + B \rightleftarrows C + D$$

$\longrightarrow$ Forward Reaction Rate = k_1 [A] [B]

$\longleftarrow$ Reverse Reaction Rate = k_2 [C] [D]

Dark Gas Phase $\rightarrow$ **Luminous Gas Phase**

In order to describe the gas phase breakdown, parameters in the dark gas phase should be used.

FIGURE 5.27 (*See color insert.*)
Reaction kinetics principle applied to the dielectric gas phase breakdown.

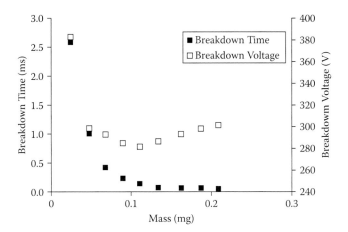

FIGURE 5.28
Breakdown time decreases with the total mass in the interelectrode system, and in the domain where the breakdown voltage increases with the mass, the breakdown occurs instantaneously. Data were obtained with the reactor shown in Figure 5.17.

phase). Accordingly, in the lower-pressure domain, the establishment of sustainable discharge current occurs with appreciable time lag as shown in Figure 5.28.

5.7 Electronegativity of Atom and Efficiency of Electron-Impact Reactions

The probability or efficiency of collision of two gases in the gas phase is generally examined as a function of the collision diameter of two spherical bodies; however, in electron-impact reactions under consideration, such a concept of cross section does not make sense because of the overwhelming difference in the diameters of colliding particles. For instance, if one picks an example of a collision between an electron and an argon atom, and assigns the electron to be a steel ball with diameter of 1 mm, then the corresponding diameter of the argon atom is about 28 m (i.e., the diameter of the argon atom is 2.8×10^4 times larger than the diameter of the electron, and the collision diameter is practically the diameter of argon).

This size difference should also be considered in the acceleration of charged particles. In particular, when an electron is accelerated to gain enough energy to ionize an argon atom by the electron-impact ionization, the ion produced has the same charge of the opposite sign, but the mass that carries the positive charge is roughly 3×10^{13} times greater than the mass of the electron. In order to accelerate Ar^+ toward the cathode, as the gas phase ionization hypothesis requires, a much higher electric field must already exist; conversely, the electric field that caused the electron impact ionization of the argon atom cannot move Ar^+ toward the cathode, which leaves the positive ion as a sitting target for recombination by a steady flux of electrons emitted by the applied electric field once the secondary electron emission occurs. The neutralization by recombination of Ar^+ and an electron leads to cessation of secondary electron emission from the cathode surface, if it did occur.

Yasuda found, in the early stages of his plasma polymerization study, the empirical rule of thumb termed the "iN/Out Rule" [24,25], which means that N atoms present in gas molecules (monomer) will be incorporated into the plasma polymer deposition, and N atoms present in the substrate polymers or polymer depositions will remain in the solid phase. Also, O atoms present in gas phase molecules will not be incorporated into the plasma deposition, and O atoms present in the substrate polymers or polymer deposition will be ablated from the solid phase. The inclusion or exclusion is not in a definitive sense but in the trend, because plasma polymerization, as well as plasma treatment, is the cumulative effects of repeating steps of activation and quenching. However, there are observable trends according to the rule.

This rule of thumb has been found to be in accordance with the electron affinity, or electronegativity, of atoms in the light of the tribo-electric series investigated from the viewpoint of the interfacial electron transfer [9,10]. Electronegativity of an atom could be paraphrased as the capability of the atom to attract and hold electrons in the vicinity of the atom. Namely, O is a highly electronegative atom (next to F), and N is the atom with the lowest electronegativity, so far as the atoms generally encounter in the chemical structure of conventional polymers and respective monomers.

The principle behind the iN/Out rule indicates that the bombardment of energetic electrons to a molecular gas that has many atoms with different electronegativities does not occur in a completely random manner; instead, electrons preferentially bombard atoms with higher electronegativity. Electrons preferentially hit O atoms in a molecule, which could lead to the splitting of O atoms from the molecule. In the case of N-containing molecules, electrons hit more electronegative portions of the molecules, yielding the minimum influence on N atom in a molecule. This trend sets the selectivity of electron bombardment.

5.8 Gas Phase Breakdown as Functions of the System Parameters

The system parameters in most experimental setups are fixed, and the only variable parameter is the system pressure. Accordingly, it is worth examining the influence of system pressure on the overall appearance of glow as

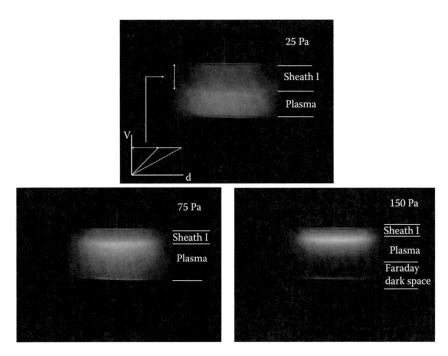

FIGURE 5.29 *(See color insert.)*
Changes of glow characteristics of DC glow discharge of argon with pressure.

a function of pressure. Figure 5.29 shows pictorial representations of glow in the reactor, shown in Figure 5.17, that show the change of glow characteristics as a function of the system pressure while all other parameters are fixed:

At *25 Pa*, which is below the pressure at the minimum breakdown voltage (transition point pressure), the cathode dark space (energy input sheath designated as Sheath-I) occupies nearly half of the interelectrode space, and the uniform intensity glow extends to the anode surface. In this case, the entire glow is the energy-dissipating sheath (designated as Sheath-II).

At *75 Pa*, at the transition point pressure, the width of Sheath-I shrinks, and the intense glow develops at the end of cathode fall, which is followed by uniform but less intense negative glow, which occupies the major portion of the interelectrode space. In this case, the energy-dissipating sheath (Sheath-II) is the negative glow/anode interface.

At *150 Pa*, which is above the transition point pressure, the width of Sheath-I shrinks further, and the width of negative glow also shrinks while its intensity increases. Faraday dark space (Sheath-II) occupies a considerable portion of the interelectrode space. In all cases shown in Figure 5.29, the width of Sheath-I, in which electrons are accelerated, decreases with the system pressure (i.e., d_L decreases with increasing pressure yielding the increased effective electric field, V/d_L). As d_L decreases with increasing pressure, the intensity of glow at the boundary of dark space increases at the expense of the extending width of negative glow, indicating that electron-impact reactions occur intensely within a narrow band of gas phase at higher electric field (V/d_L) in luminous gas phase.

In the lower-pressure domain, the breakdown current establishes with a finite delay time as depicted in Figure 5.28. The time lag of the gas phase breakdown depends on the mean free-path length of gas, which depends on the mass in the interelectrode volume (M_v), because the breakdown process is the mass action between electrons and gas particles. As the mass reaches the critical value, the establishment of broken-down current becomes virtually instantaneous, because the sufficient amount of mass (shorter mean-free path) is available in the gas phase.

Breakdown voltage, V_b, and breakdown current, I_b, cannot be directly measured in RF discharge, and breakdown power, W_b, is the experimental parameter. Therefore, the breakdown process is investigated with both V_b and I_b, and results are also presented in terms of all derivable parameters from V_b and I_b as shown below. $W_b = V_b * I_b$ (conductivity $= I_b/V_b$). Input energy per mass (Yasuda parameter) is $W_b/F*M = W_b/M_v$, where F is volume or molar flow rate, M is the molecular weight of gas, and M_v is total mass in glow volume in unit time.

Although all y-axis values should be plotted against M_v for general cases, the system pressure, p, is used to more easily grasp general trends, because most experimental works are carried out in a fixed reactor, in which other parameters that influence the breakdown process are not experimental variables, and $p = a * M_v$, where a is proportionality constant for a given system that can be calculated for each system. All data for breakdown processes are obtained by using a bell jar–type reactor, which is used for the MLCVD of CH_4. Magnets for the magnetron are removed for investigation of gas phase breakdown without influence of a magnetic field.

Figure 5.30 [26] depicts breakdown voltage as functions of pressure for Ar, N_2, O_2, and CH_4. The plots tell us what we could extract from the breakdown voltage versus pressure plots:

1. Breakdown voltage versus pressure plots for the four gases, which represent (a) mono-atomic non-deposition, non-dissociable gas, Ar, (b) non-deposition molecular gas, N_2 and O_2, and (c) deposition molecular gas, CH_4, are nearly the same.
2. The breakdown voltage of Ar, which has been the main topic of breakdown phenomena, is the lowest among the gases investigated.

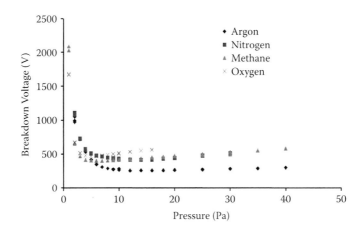

FIGURE 5.30
The dependence of breakdown voltage on system pressure for Ar, N_2, O_2, and CH_4.

3. There are two clearly identifiable domains of the breakdown gas phase separated by a transition point pressure, p_T.
4. The transition point pressure, p_T, is nearly independent of type of gas.

The marginal dependence of breakdown voltage on types of gas might give an erroneous conclusion that the breakdown process is nearly independent of the types of gas, which provides the justification for extending the results obtained with Ar discharge to discharge of various gases. This is the key point that the breakdown voltage cannot be used to investigate the dielectric breakdown of the gas phase in general, based on the concept of gas phase breakdown as described previously. The key issue becomes very clear when one examines the breakdown current at the onset of the gas phase breakdown as shown below.

Figure 5.31 [26] shows the dependence of breakdown current on system pressure. In contrast to the breakdown voltage, the breakdown current shows great dependency on the type of gas. It is also quite evident that breakdown current increases as the system pressure increases only above the transition point pressure for all gases. Below the transition point pressure, breakdown current is very low and nearly constant, regardless of values of breakdown voltage, which indicate the significant feature of the transition point pressure.

Based on these assessments of V_b and I_b, the phase diagram of gas phase breakdown can be drawn to characterize the dielectric breakdown of gas phase in DC discharge. Figure 5.32 [26] depicts the phase diagram using the plot obtained with Ar. In domain [I], the dielectric breakdown of the gas phase cannot occur. In domain [II], breakdown voltage decreases with pressure, but breakdown current does not vary with pressure and remains

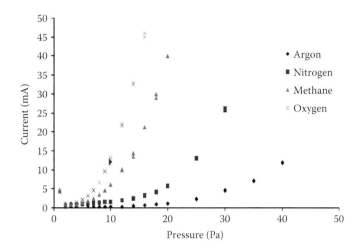

FIGURE 5.31
The breakdown current as functions of system pressure for Ar, N_2, O_2, and CH_4.

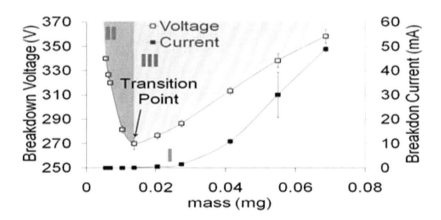

Phase Diagram of Dielectric Gas Phase Breakdown

I   Dielectric gas phase
II   High Voltage/Low Current
III   Low Voltage/High Current

FIGURE 5.32
Phase diagram of dielectric breakdown of gas phases depicted by breakdown voltage and breakdown current: Phase [I]; nonconducting gas phase, Phase [II]; conducting gas phase (high voltage/low current), Phase [III] conducting gas phase (low-voltage/high current).

at a very low value. Domain [II] can be characterized by high-voltage/low-current discharge. In domain [III], breakdown voltage increases slightly with pressure, but current increases sharply with pressure, though its rate depends on type of gas. Domain [III] can be characterized by low-voltage/high-current discharge.

Mono-atomic (nondeposition) gas, Ar, shows a very small increase of current as pressure increases, indicating the ionization is least dependent on the system pressure. Because the dependence of breakdown voltage on the system pressure is small, all derived parameters show a similar trend observed with breakdown current, as shown in Figure 5.33 [26] for conductivity of the broken-down gas phase, in Figure 5.34 [26] for breakdown wattage, and in Figure 5.35 [26] for energy input per unit mass in J/kg.

These figures clearly show that the breakdown phenomenon investigated with breakdown voltage of Ar has very little relevance in characterizing the breakdown phenomena of various types of gases, particularly of reactive molecular gases and molecular deposition gases. The figures for energy per mass of gas indicate that O_2 and CH_4 are capable of transferring significantly greater energy per mass of gas than N_2 and Ar. Here again, the difference of types of gas are quite evident, particularly differences from Ar. W/M_v is essentially identical to W/FM, developed to normalize the operational parameter of RF plasma polymerization in different reactors, which has been known as the *Yasuda Parameter* among practitioners of plasma polymerization avoiding such vague nonscientific terms as "high power," "low power," "high pressure," and "low pressure." Figure 5.35 clearly demonstrates that W/FM or W/M_v is the key parameter for plasma polymerization and plasma treatment with reactive gases when the current

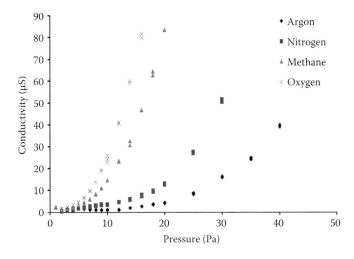

FIGURE 5.33
Conductivity of broken-down gas phase as a function of pressure for Ar, N_2, O_2, and CH_4.

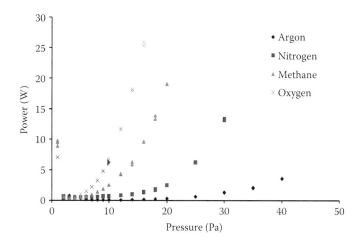

FIGURE 5.34
Dependency of breakdown power on pressure for Ar, N_2, O_2, and CH_4.

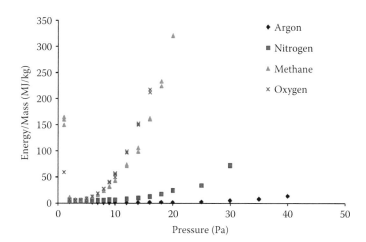

FIGURE 5.35
Dependency of energy per unit mass, $W/FM = W/M_v$, on pressure for Ar, N_2, O_2, and CH_4.

value is not available (i.e., in RF discharge). Those data clearly show that Ar discharge is significantly different from the rest of gases, and fundamental data obtained with Ar DC discharge have little value in DC discharge of reactive and deposition gases and also the breakdown phenomena in RF discharge. It is worth noting that the difference of electronegativity of O_2 and N_2 is clearly visible in the breakdown current, while the breakdown voltage difference is less pronounced.

References

1. Yasuda, H., *Luminous Chemical Vapor Deposition and Interface Engineering*, Marcel Dekker, New York, 2004.
2. Yasuda, H. and Q. Yu, *Plasma Chemistry & Plasma Process*, 24, 325, 2004.
3 Yu, Q., C. Huang, and H. Yasuda, *J. Polym. Sci., A: Polym. Chem.*, 42, 1042, 2004.
4 Yasuda, H., and Q. Yu, *J. Vac. Sci. Technol.* A 22(3), 472, 2004.
5 Yasuda, H., *Plasma Processes and Polymers*, 4, 347, 2007.
6 Yasuda, H., L. Ledernez, F. Olcaytug, and G. Urban, *Pure and Applied Chemistry*, 80, No. 9, 1883, 2008.
7. Chapman, B., *Glow Discharge Processes*, John Wiley, New York, NY, 1980.
8. Brown, S. C., *Introduction to Electrical Discharges in Gases*, John Wiley, New York, NY, 1966.
9. Yasuda, H., J. Charlson, E. Charlson, T. Yasuda, M. Miyama, and T. Okuno, Langmuir, 7, 2394, 1991.
10. Charlson, E., J. Charlson, J. Burkett, and H. Yasuda, *IEEE Trans. on Electrical Insulation*, 27, No. 6, 1136, 1992.
11. Paschen, F., *Annalen der Physik*, 273 (5), 69–96, 1889.
12. Townsend, J. S. E., *Electricity in Gases*, Oxford, Clarendon Press, 1915.
13. Cobine, J. D., *Gaseous Conductors: Theory and Engineering Applications*, Dover Publications, 1958.
14. Sato, S., Y. Kuboyama, M. Sone, and H. Mitsui, *Annual Report, Conference on Electrical Insulation and Dielectric Phenomena*, 292, 1992.
15. Osmokrovic, P., *Plasma Science, IEEE Transaction* 21(6), 1993.
16. Auday, G., Ph. Guillot, J. Galy, and H. Brunet, *J. Appl. Phys.* 83, 5917, 1998.
17. Hartman, P., Z. Donko, G. Bano, L. Szalai, and K. Rozsa, *Plasma Sources Sci. Technology*, 9, 183, 2000.
18. Lisovskii, V. A. and S. D. Yakovin, Springer, 45, 727, 2000.
19. Pejovic, M. M., G. S. Ristic, and J. P. Karamarkovie, *J. Phys. D: Appl. Phys.*, 35, 91, 2002.
20. Hassouba, M. A., F. F. Elaskshar, and A. A. Garamoon, *Fizika*, A 11, 2, 81, 2002.
21. Osmokrovic, P. and A. Vasic, *IEEE Transaction on Plasma Science*, 33, 1672, 2005.
22. Kobayashi, H., A. T. Bell, and M. Shen, *J. Macromol. Sci. Chem.* 10, 491, 1976.
23. Yu, Q. and H. Yasuda, *Plasmas & Polymers*, 7, 415, 2002.
24. Yasuda, T., M. Gazicki, and H. Yasuda, *J. Appl. Polym. Sci., Appl. Polym. Sym.*, 38, 201, 1984.
25. Yasuda, H., and T. Yasuda, *J. Polym. Sci., Polym. Chem. Ed.*, 38, 943–953, 2000.
26. Yasuda, H., L. Ledernez, and F. Olcaytug, Data to be published, 2010.

6

Influence of Magnetic Field on Luminous Gas Phase

6.1 Influence of Magnetic Field on Roles of Electrons

6.1.1 Changes in Distribution Profile of T_e and N_e in Luminous Gas Phase

A magnetic field superimposed on an electrode surface used in direct current (DC) discharge changes the distribution profiles of electron temperature, T_e, and electron density, N_e, which means that the electric discharge with magnetic field is different from that without a magnetic field [1,2]. Profiles for DC discharge of Ar without magnetic field (a plain cathode plate against a plain anode plate) are shown in Figure 6.1. Electron temperature reaches the maximum value and starts to decrease sharply with further increase of distance from the cathode surface. The maximum point corresponds to where the intense glow develops at the edge of the cathode fall dark space. Electron density, on the other hand, is low, while electrons are being accelerated, and the product of $T_e * N_e$ remains nearly constant in the entire domain where measurements are made, which is reasonable because the energy transferred to the gas phase is a constant. Those values are, of course, dependent on the operation factors that influence the breakdown process (i.e., separation distance, surface area of electrode, system pressure, and type of gas as described in Chapter 5).

Electron temperature rises in the cathode dark place (the energy input sheath) as the distance from the cathode surface increases. This is what we anticipate, because electrons are accelerated in the dark space by the applied electric field, which is an important part of the broken-down gas phase, although one might consider that the dark space is not in "plasma state" based on a narrow view or definition of "plasma state." A Langmuir probe measures voltage-current in a broken-down electrically conducting gas phase, regardless of whether it is a luminous part or a nonluminous part of the broken-down conducting gas phase.

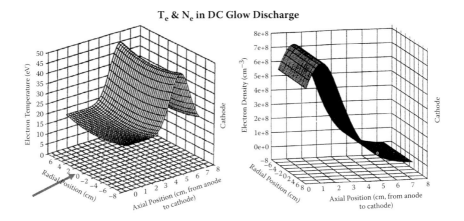

FIGURE 6.1

Distribution profile of electron temperature, T_e, and electron density, N_e, in DC discharge without magnetic field.

Distribution profiles of T_e and N_e by Langmuir probe measurement tell us the following:

1. T_e and N_e are measured in the dark space as well as in the glow, once the gas phase is dielectrically broken down.
2. The product of T_e times N_e is nearly constant, indicating that the energy input into the luminous gas phase is a constant.
3. There is no indication that the value of T_e or N_e could be correlated to the glow characteristics, with respect to the location of luminosity.
4. In a large portion of the glow, T_e is less than the value of ionization energy of Ar.
5. In view of the increasing N_e near the anode, it is unreasonable to assume the electroneutrality in low-pressure "plasma."
6. The dielectrically broken-down gas phase, as a whole, cannot be represented by "nonequilibrium plasma" in a context that nonequilibrium refers to energy disparity only.

The luminous gas phase created by low-pressure glow discharge of argon is not homogenous, and the homogeneity could be applicable only within relatively thin layers parallel to the electrode surface in the luminous gas phase. The whole luminous gas phase could be represented by the sum of such a thin layer resembling an onion-layered structure rather than a uniform gas phase. The influence of the magnetic field, described below, is investigated using identical conditions, except for the presence of a magnetic field either on the cathode or on the anode.

6.1.2 Magnetron Cathode (MC) and Nonmagnetron Anode (NA)

The magnetic field superimposed on the cathode surface changes the glow near the cathode surface completely, and narrow intense glow develops in the center portion of the cathode surface forming a toroidal glow with the concentric circular arrangement of the magnetic field as shown in Figure 6.2. The toroidal glow, shown in Figure 6.3, develops near the surface but in the gas phase. The glow shown is one half of the 15 kHz discharge, which represents the cathodic glow in DC discharge. The ionization glow of Ar does not cover the entire surface of the cathode, and the intense glow remains only in the toroidal glow. When the toroidal glow develops, the glow generally recognized as the negative glow, which fills the majority of space of luminous gas phase, is fairly faint. The deposition of material in magneto-luminous chemical vapor deposition (MLCVD) occurs in this faint glow (i.e., dissociation products caused in the toroidal glow to move into the faint glow domain). The substrates move in and out of this domain, in the center part of two toroidal glows. The movement of substrate averages out the deposition on the substrate surface.

Figure 6.4 schematically depicts the location of toroidal glow with respect to the profile of the magnetic field for the MC/NA discharge mode. The distribution profiles of T_e and N_e for MC/NA mode are shown in Figure 6.5.

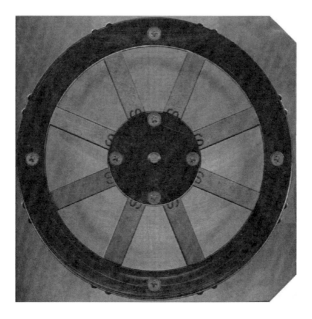

FIGURE 6.2
Circular arrangement of eight permanent magnets bridging the center iron plate and a circular iron ring concentrically placed behind the cathode plate.

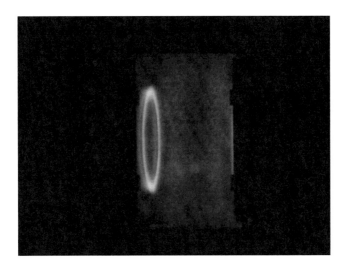

FIGURE 6.3
The toroidal glow develops near the cathode surface.

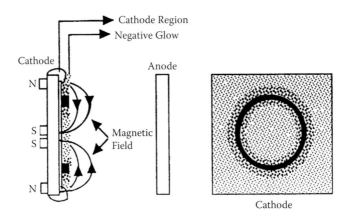

FIGURE 6.4
The toroidal glow in a MC/NA arrangement.

The toroidal glow develops in the gas phase near the cathode surface, where the electric field and magnetic field are orthogonal, and consists of a large number of low-energy electrons. The highest value of T_e is observed at the center of the electrode, which is the dark hole of toroidal glow. The value of N_e is very low at the center part, and the glow is obscured by the much more intense toroidal glow. How such an intense toroidal glow develops with very low-energy electrons seems to be the key factor of the magnetron discharge, which is discussed in Section 6.3.

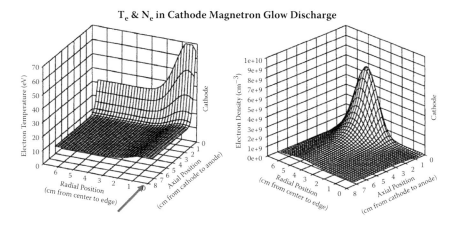

FIGURE 6.5
Distribution profiles of electron temperature, T_e, and electron density, N_e, in an MC/NA arrangement of DC discharge. Toroidal glow develops with a large number of low-energy electrons.

6.1.3 Nonmagnetron Cathode (NC) and Magnetron Anode (MA)

Figure 6.6 schematically depicts the shape of glow with respect to the shape of magnetic field in NC/MA mode. Figure 6.7 depicts the distribution of electron temperature and electron density in the discharge of argon with a plain metal plate cathode and two magnetron anodes. The distribution profile of electron temperature is similar to that observed without magnetron anode, shown in Figure 6.1. The conspicuous difference is the absence of the edge effect, which is caused by the influence of the magnetic field on the

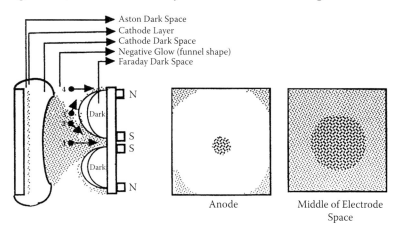

FIGURE 6.6
Glow develops in a MC/NA arrangement of DC discharge. There is no toroidal glow and shaped negative glow near the magnetron anode.

T_e & N_e in Anode Magnetron Glow Discharge

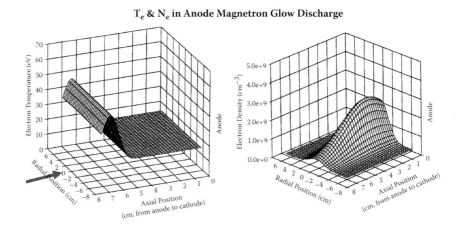

FIGURE 6.7

Distribution profiles of electron temperature, T_e, and electron density, N_e, in a MC/NA arrange-ment of DC discharge. The electron temperature profile resembles that of NC/NA mode dis-charge shown in Figure 10.1, but the electron density profile differs significantly with the large peak pointing to the center of the anode surface.

anode. Electrons are pulled toward the center of the anode where the hole of the donut-shaped magnetic field exists and causes a slightly curved ioniza-tion glow near the cathode.

The edge effect, the higher electric field concentration on the edge of the electrode, can be eliminated by the use of an anode magnetron. The major effect of anode magnetron is seen in the electron-density profile. The highest peak of electron density occurs at the center of the magnetron anode, which corresponds to the center hole of the magnetic field. A large number of low-energy electrons concentrate in the center part of the anode surface. This concentration of low-energy electrons has very important implications on the characteristics of the luminous gas phase, as described below.

6.2 Shaping of Negative Glow Near the Magnetron Anode

6.2.1 Negative Glow of Argon

Figure 5.2 (in Chapter 5) shows the glow of Ar at the inception of glow dis-charge created by pairing a plain plate cathode with a magnetron anode, NC/MA mode [3–7]. The ionization glow of Ar appears approximately 2 cm away from the cathode surface, under the conditions employed, which coincides with the location of the maximum peak of electron temperature in the distribution profile. The conspicuous influence of the magnetic field on the nature of the luminous gas phase can be seen first on the shape of the negative glow near the anode surface. Without a magnetic field on the anode, the negative glow more or less uniformly extends toward the anode surface. With a magnetic field on the anode, the negative glow of Ar is shaped following the path of electrons as depicted in Figure 6.6.

6.2.1.1 Reexcitation of Photo-Emitting Species by Low-Energy Electrons

Why the negative glow is shaped by the magnetic field has significant implications for elucidation of the luminous gas phase. The following important factors should be considered:

1. The magnetic field applied on the anode surface is too weak to influence the movement of species created by ionization of Ar but strong enough to influence the path of electrons because of its extremely small mass.
2. Electrons are not photon-emitting species.
3. The electric field as well as the magnetic field have no influence on the movement of photon-emitting neutral species created by ionization of the Ar atom, which moves randomly in the luminous gas phase.
4. A photon-emitting excited species eventually loses the energy corresponding to the photo-emission energy upon emitting the photon and becomes a non-photon-emitting pseudo-excited species.
5. The energy loss due to the photoemission, however, is only a fraction of the excitation energy to create the photon-emitting excited species from the ground state of the Ar atom, as schematically depicted in Figure 5.4 (Chapter 5).
6. There are large numbers of electrons with electron energy higher than the energy necessary to reexcite those pseudo-excited species. (See Figure 6.1 that shows electron energy and electron density distribution profiles.)
7. The electrons near the anode with magnetic field has the highest peak at the center of the electrode, but the peak width is narrow, and the number of electrons is very low beyond the peak (see Figure 6.7).

If photo-emitting excited species, created in the ionization glow, at the edge of cathode fall region, reached the negative glow region near the anode before emitting photons and then emit photons (i.e., without reexcitation), the shaping of the negative glow should not occur. The excited species are neutral, and their movement is the random gas movement that will not be influenced by either electric or magnetic field. The shaping of negative glow means that the low-energy electrons accumulating near the center of the anode are involved in the reexcitation of the pseudo-excited neutral species, as depicted in Figure 5.4 (Chapter 5), near the anode surface. The excitation of the Ar atom near the anode surface is excluded from consideration, because electron energy in the domain is too low to excite Ar atoms.

The shaping of negative glow near the magnetron anode surface indicates that the photo-emitting species that have lost energy by emitting photons are repeatedly reexcited by the low-energy electrons in the region away from the energy input sheath. In other words, the reexcitation of photo-emitting neutral species by low-energy electrons occurs throughout the luminous gas phase but cannot occur at the location where electrons do not exist or electron energy is too low. Because most electrons are pulled into the center portion of the anode by the magnetic field, the reexcitation occurs mainly at the center part of the magnetron anode.

This explanation also implies that the excited species Ar^*, not Ar^+, are the major species in the luminous gas phase created by the electron impact on Ar atoms (i.e., the glow of Ar discharge is more appropriately expressed as "luminous gas phase" rather than "ionized gas phase"). The ionization of Ar atoms seems to be the major path for creating photo-emitting neutral species via recombination of Ar ions and electrons in view of the fact that there is no discernible glow at the corresponding energy level of excitation energy, unlike the clear existence of the dissociation glow with deposition gases, which is separated by an identifiable dark space in between the dissociation and ionization glows.

6.2.2 Influence of Magnetron Anode on Glow Characteristics of Deposition Gas

An anode magnetron also influences characteristics of the luminous gas phase of deposition gas. Figure 6.8 depicts the dissociation glow and negative glow of trimethylsilane (TMS) without magnetic field, a plain cathode at the center against two plain anodes. This glow provides the reference for comparing the influence of glow with magnetic field on anode shown in Figure 6.9, which shows the same luminous gas phase created with a magnetic field, a plain cathode against two magnetron anodes. Without a magnetic field, the shaping of negative glow is absent. The shaping of negative glow by the magnetron anode observed with Ar discharge is also clearly present with the deposition gas. The magnetic field of the anode pushes both dissociation glow and ionization glow toward the cathode surface; the thickness of dissociation glow becomes slightly thinner, and the ionization glow becomes slightly concave toward the anode as seen in the case of Ar discharge.

Not all negative glows of organic molecules are shaped by the magnetic field on the anode, as shown in Figure 6.10. Most F-containing molecular gases show no influence of magnetic field shaping of negative glow as

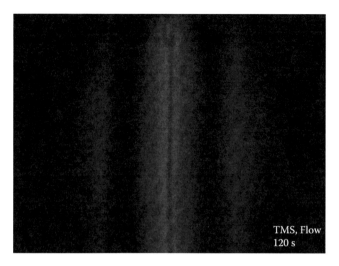

FIGURE 6.8 (*See color insert.*)
Dissociation glow and ionization glow of TMS with two plain anodes without magnetic field.

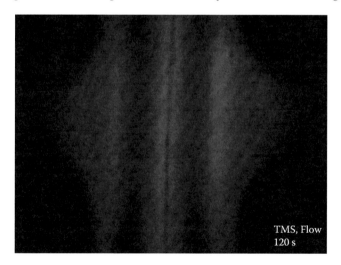

FIGURE 6.9 (*See color insert.*)
Dissociation glow and ionization glow of TMS with two anodes with magnetic field.

shown in Figure 6.11. It is likely that electron attachment to highly electro-negative F atoms yields dissociated species with negative charge, which follows the electric field (i.e., photo-emitting species are pulled to the anode). Furthermore, when electron attachment to a dissociation gas occurs, the number of electrons in luminous gas phase decreases compared to the case in which electrons mainly transfer energy to colliding gas. Probably reflecting the electron attachment, the dissociation glows of gases shown in Figure 6.11 seem to be weaker than glows of gases without F in the molecule.

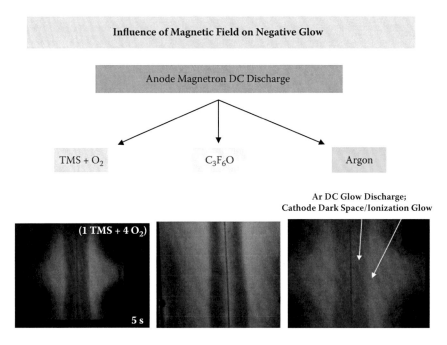

FIGURE 6.10 (*See color insert.*)
Magnetic shaping of negative glow of various gases in DC discharge. Not all gases show the effect of magnetic field. Dissociation glow and ionization glow of TMS with two anodes with magnetic field.

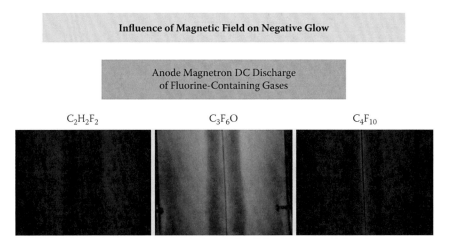

FIGURE 6.11 (*See color insert.*)
Negative glow of fluorine-containing gases in anode magnetron DC discharge.

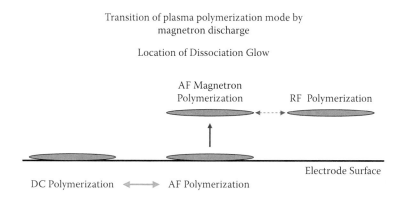

FIGURE 6.12
The location of dissociation glow: the interrelationship among DC plasma polymerization, AF plasma polymerization, AF magnetron plasma polymerization, and RF plasma polymerization.

6.2.3 Shift of Dissociation Glow from Cathode Surface to Gas Phase

The dissociation glow of organic molecules in DC discharge covers the entire surface of the cathode in the absence of a superimposed magnetic field. In the presence of a magnetic field, however, the shape of glow changes drastically as described above, and the glow concentrates in the toroidal glow, leaving the rest of the cathode surface in the dark. On this transformation, the dissociation glow in the form of toroidal glow is also lifted up from the cathode surface to the gas phase (i.e., the toroidal glow is not in contact with the cathode surface as can be seen in Figure 6.3). Figure 6.12 schematically depicts the location of the dissociation glow as the frequency of the power source increases. The lifting of the dissociation glow from the cathode surface to the gas phase has very important implications in the sustainability of plasma polymerization, which is described in some detail in Chapter 8. Table 6.1 summarizes the major differences.

6.3 Influence of Magnetic Field on Dielectric Breakdown of Gas Phase

The most significant influence of the magnetic field is seen in the dielectric breakdown process. The distribution profiles of T_e and N_e change significantly by the influence of the magnetic field as shown in the above sections, indicating the mechanism of the dielectric breakdown of gas with superimposed magnetic field is completely different from that without magnetic field. The influence of magnetic field on the dielectric breakdown of the gas phase in low pressure is so astonishingly great that the

TABLE 6.1

Comparison of Audio Frequency (AF) Plasma Polymerization and Magnetron-AF Plasma Polymerization

	AF Plasma Polymerization	Magnetron-AF Plasma Polymerization
Energy input parameter	I/FM by current control mode operation	I/FM by current control mode operation
Creation of polymerizing species	Impact of low-energy electrons in the linear motion toward anode	Impact of low-energy electrons circling along the magnetic field line
Location of dissociation	Dissociation glow is on the cathode surface	The toroidal dissociation glow in gas phase near the surface of electrodes
Confinement of glow	No	Yes
Yield of deposition on substrate	Low; most deposition on the cathode surface	High; substrate movement in and out of glow is essential
Deposition on the cathode surface	High; deposition occurs on the whole surface	Low; a part of circular deposition on the cathode surface has very little to no deposition
Wall contamination by plasma polymer deposition	Moderate	Very little
Change of plasma polymerization behavior with reaction time	Considerable extent	Negligible after a short break-in time
Continuous long-time operation	Difficult	Easy
Scale-up of process	Difficult to very difficult	Easy

breakdown should occur by a completely different mechanism. The influence of the magnetic field on the breakdown phenomena could be visualized by the following four figures, which compare the influence of the type of gas, Ar, N_2, O_2, and CH_4, on the pressure dependence of the breakdown voltage and breakdown current. Figure 6.13 [11] depicts the comparison of breakdown voltages for the four gases without magnetic field. Figure 6.14 [11] compares the same with magnetic field, but the scale of the y-axis is expanded and covers a narrower range of voltage in order to see the influence of types of gases. The effect of the magnetic field can be seen by comparing both figures.

The influence of type of gas on breakdown voltage without magnetic field is relatively small, but there are appreciable differences due to the types of gas:

O_2 *and* CH_4: Breakdown voltages below the transition point pressure decrease, approaching the transition point, and level off at nearly 100 V lower than the breakdown voltage at the transition point voltages without magnetic field (i.e., the sharp increase of breakdown voltage below the transition point pressure does not occur).

	Rear end Approach		Front end Approach	
	LBL Coating		**LCVD Coating**	
Initial Cost of Coater	$250,000		$3,000,000	
Annual Operation Cost				
	Quantity (Kg)	**Cost (US$)**	**Quantity (Kg)**	**Cost (US$)**
Coating Materials	**7,759,257.4**	**$181,277**	**1.9**	**$146**
Coating Yield	0.05%		13.59%	
Total Effluent of Processes	374,400		~ 0.2	
Effluent Treatment Cost		$5,971,292		$0
Depreciation		$25,000		$300,000
Total Cost (without labor)		**$6,177,569**		**$300,146**

COLOR FIGURE 3.4
Comparison of operation costs for conventional wet chemistry process and LCVD coating and impact of environmental remediation cost.

Ar DC Glow Discharge; Cathode Dark Space/Ionization Glow

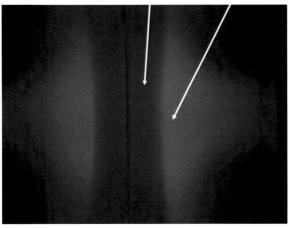

COLOR FIGURE 5.2
Glow develops with Ar DC discharge, with anode magnetrons.

DC Glow Discharge of TMS
Cathode Glow/Dark Space/Ionization Glow

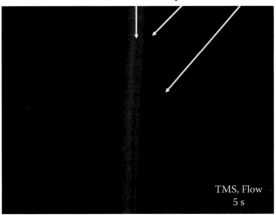

TMS, Flow
5 s

COLOR FIGURE 5.5
Glow develops with TMS DC glow discharge, with anode magnetrons.

Change of Dissociation Glow & Ionization Glow with Reaction Time

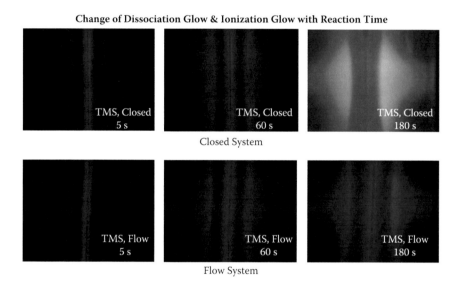

Closed System

Flow System

COLOR FIGURE 5.8
Change of dissociation glow and ionization glow of TMS with reaction time.

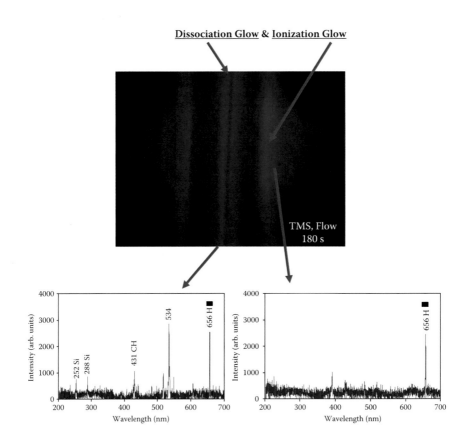

COLOR FIGURE 5.11
Major photon-emitting species in dissociation glow and ionization glow of TMS.

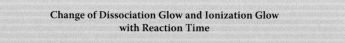

**Change of Dissociation Glow and Ionization Glow
with Reaction Time**

At the onset of Glow Discharge   After 60 seconds

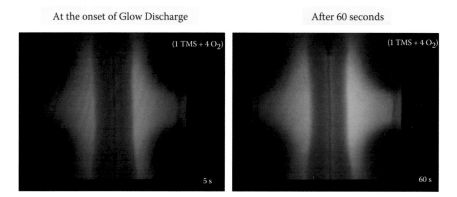

COLOR FIGURE 5.12
Change of dissociation and ionization glows due to the interaction of reactive species in dissociation and ionization glow. The ionization glow of O_2 changes to that of H_2, which is one of the dissociation products of TMS.

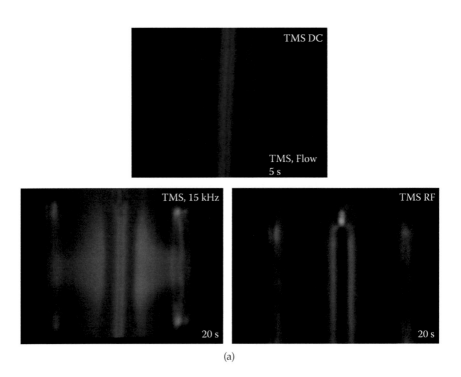

(a)

COLOR FIGURE 5.13
(a) Effect of frequency of power source on the dissociation glow of TMS.

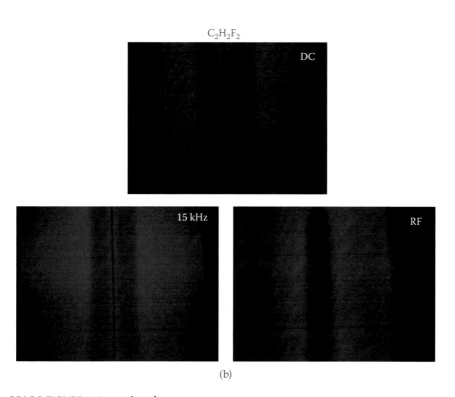

(b)

COLOR FIGURE 5.13 (continued).
(b) Effect of frequency of power source on the dissociation glow of $C_2H_2F_2$.

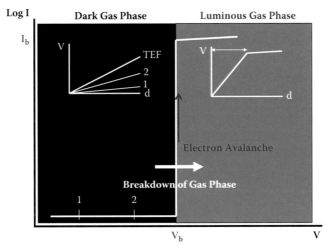

Dielectric Breakdown of Gas Phase

COLOR FIGURE 5.19
The dielectric gas phase breakdown in terms of applied voltage and observed electric current: *V–d* plots (insert) show the change of electric field, which is the slope; *V/d*, where *d* is the separation distance of the cathode and anode.

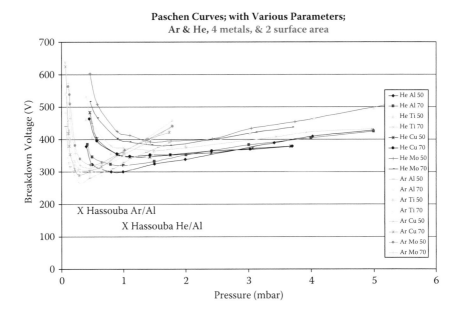

COLOR FIGURE 5.25
Paschen plots at a fixed *d*, two gases, four metals, and two surface areas.

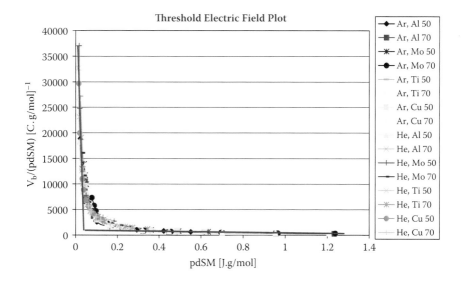

COLOR FIGURE 5.26
Threshold electric field plots of data shown in Figure 5.25.

$$A + B \rightleftarrows C + D$$

⟶ Forward Reaction Rate = k_1 [**A**] [**B**]

⟵ Reverse Reaction Rate = k_2 [**C**] [**D**]

Dark Gas Phase ➤ **Luminous Gas Phase**

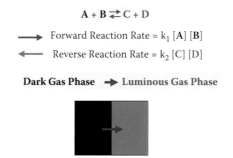

In order to describe the gas phase breakdown,
parameters in the dark gas phase should be used.

COLOR FIGURE 5.27
Reaction kinetics principle applied to the dielectric gas phase breakdown.

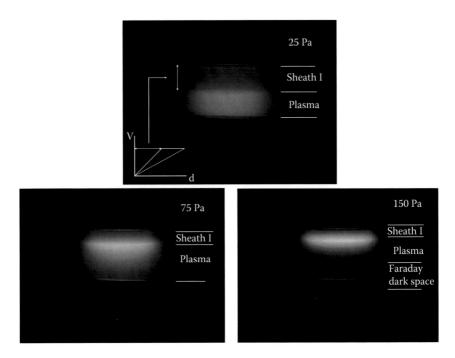

COLOR FIGURE 5.29
Changes of glow characteristics of DC glow discharge of argon with pressure.

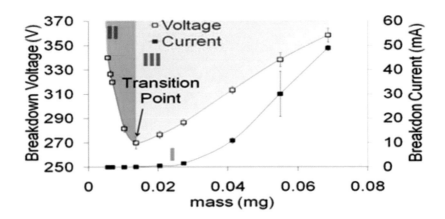

FIGURE 5.32
Phase diagram of dielectric breakdown of gas phases depicted by breakdown voltage and breakdown current: Phase [I]; nonconducting gas phase, Phase [II]; conducting gas phase (high voltage/low current), Phase [III] conducting gas phase (low-voltage/high current).

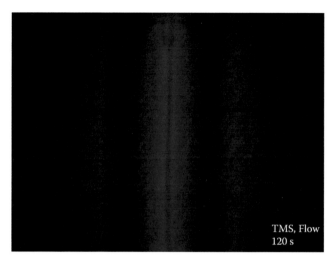

COLOR FIGURE 6.8
Dissociation glow and ionization glow of TMS with two plain anodes without magnetic field.

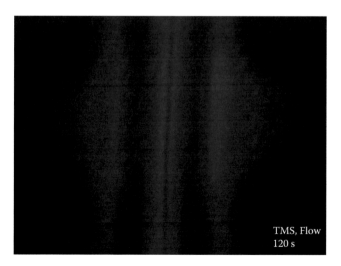

COLOR FIGURE 6.9
Dissociation glow and ionization glow of TMS with two anodes with magnetic field

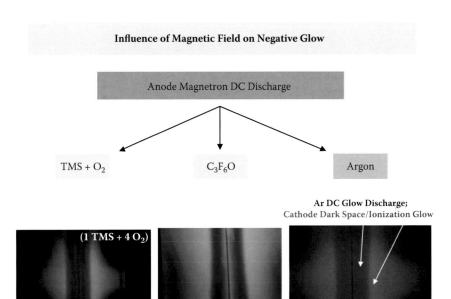

COLOR FIGURE 6.10
Magnetic shaping of negative glow of various gases in DC discharge. Not all gases show the effect of magnetic field. Dissociation glow and ionization glow of TMS with two anodes with magnetic field.

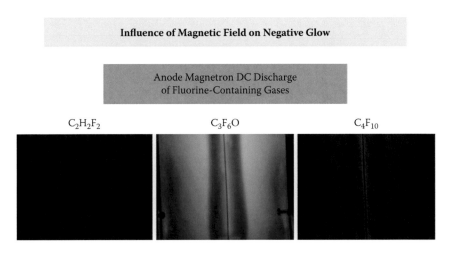

COLOR FIGURE 6.11
Negative glow of fluorine containing gases in anode magnetron DC discharge.

Magnetron Plasma Polymerization

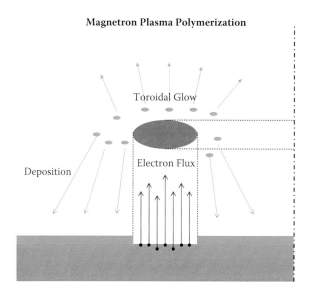

Toroidal Glow

Electron Flux

Deposition

Magnetron Plasma Sputtering

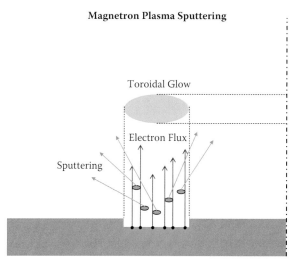

Toroidal Glow

Electron Flux

Sputtering

COLOR FIGURE 6.22
Magnetron plasma polymerization (left) and magnetron plasma sputtering.

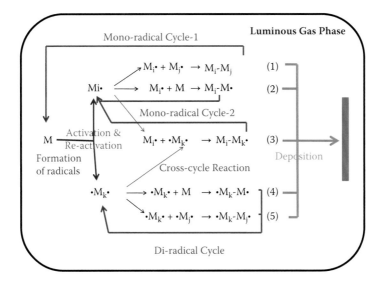

COLOR FIGURE 7.1
Polymer formation mechanism in luminous gas phase: RSGP mechanism.

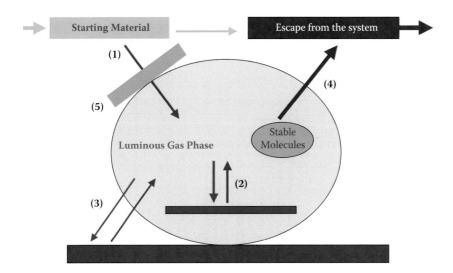

COLOR FIGURE 7.4
The CAP principle: (1) dissociation (ablation) of the monomer to form reactive species; (2) deposition of plasma polymer and ablation of solid, including plasma polymer deposition; (3) deposition to and ablation from nonsubstrate surfaces; (4) removal of stable molecules from the system; and (5) the zone in which dissociation of monomer occurs.

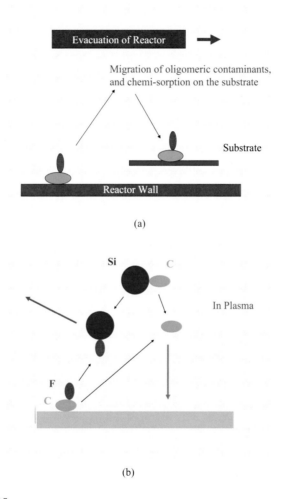

COLOR FIGURE 7.5
How F-containing contaminant interferes with plasma polymerization of TMS: (a) migration of
F-containing oligomers; and (b) the interference of TMS deposition by F-containing moieties.

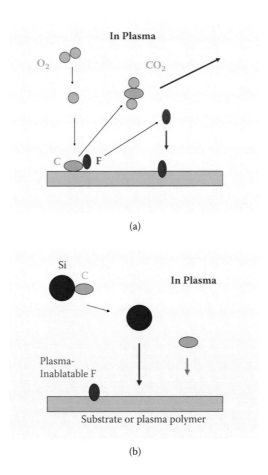

COLOR FIGURE 7.6
How oxygen plasma treatment prevents the interference of plasma polymerization of TMS by
F-containing oligomers: (a) oxygen plasma treatment; and (b) plasma polymerization of TMS.

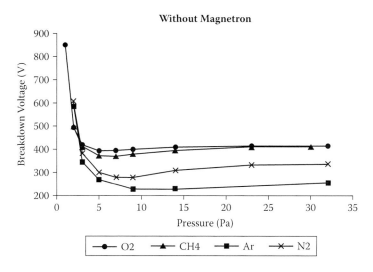

FIGURE 6.13
Breakdown voltage as a function of pressure without magnetic field for Ar, N_2, O_2, and CH_4.

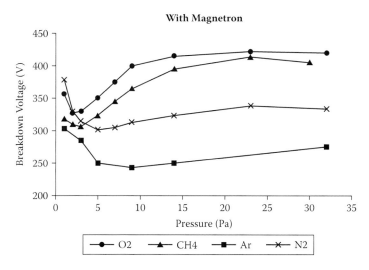

FIGURE 6.14
Breakdown voltage as a function of pressure with magnetic field for Ar, N_2, O_2, and CH_4.

N_2 *and Ar*: Breakdown voltages above the transition point pressure are similar with and without the magnetic field. The influence of magnetic field on the breakdown voltage is mainly in the domain below the transition point pressure. The sharp increases of breakdown voltage below the transition point pressure, without magnetic field, virtually disappear, and breakdown voltages below the transition point pressure are significantly lower. The increases of breakdown

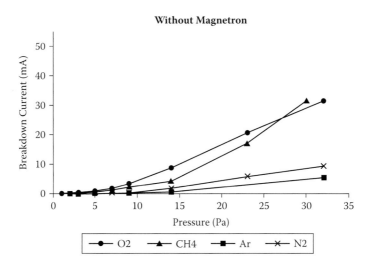

FIGURE 6.15
Breakdown current as a function of pressure without magnetic field for Ar, N_2, O_2, and CH_4.

voltages below the transition point pressure are significantly less than those without magnetic field.

Figure 6.15 and Figure 6.16 [11] are similar displays to show the influence of the magnetic field on breakdown current. As can be seen from these figures, the major influences of magnetic field on the gas phase breakdown are seen in the great increase of breakdown current:

O_2 *and* CH_4: At the lowest pressure, breakdown currents reached the ceiling current limit of the power source used. The breakdown currents are greater than the sealing current value shown. There is no transition point breakdown current. The breakdown current of CH_4 starts to decrease at higher pressure above 23 Pa.

N_2 *and Ar*: The major influence of the magnetic field seems to be limited in the domain below the transition point pressure. At the lowest pressure, breakdown currents reached the ceiling limit of current also, but breakdown currents decrease with increasing pressure, reach the minimum value at around the transition point pressure, and then follow the pattern of breakdown current observed without a magnetic field, although the breakdown current is significantly higher than that without magnetic field. The minimum breakdown currents of N_2 are very low, nearly the same as values of breakdown currents below the transition point pressure without a magnetic field.

For all gases examined, the breakdown current at the lowest pressure of measurements, where very low current (less than 1 mA) is measured without magnetic field, is over 50 mA, which is the ceiling current limit of the

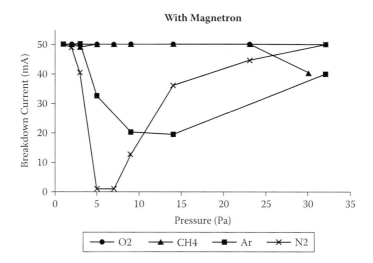

FIGURE 6.16
Breakdown current as a function of pressure with magnetic field for Ar, N_2, O_2, and CH_4.

power supply used. These high currents are created by the lower breakdown voltages than those without magnetic field. Even with Ar and N_2, for which the influence of magnetic field is less pronounced among gases examined, breakdown current reaches the ceiling current of the power supply. It is a very interesting question, what would be the breakdown current in the lower end of the pressure scale for O_2 and CH_4, if the power supply could deliver higher current?

The results shown in these four figures are significantly different from what we might anticipate based on the conventional wisdom of the gas phase breakdown, because the conventional concept does not consider the importance of current associated with the gas phase breakdown. It is difficult to comprehend what is happening in the gas phase breakdown, unless we recognize the major factor of the gas phase breakdown is the creation of gas phase species that allow transfer of energy in the form of current. It is important to recognize that O_2 that is a highly electronegative molecule yields much higher breakdown current than N_2 that is the least electronegative molecule examined. The dissociable gas CH_4 also yields much higher breakdown current than nondissociable gas Ar.

Those figures show clearly distinguishable behaviors of two groups: one group consisting of highly electronegative molecular gas O_2 and highly dissociable deposition gas CH_4, and another group consisting of less electronegative molecular gas N_2 and nondissociable mono-atomic gas Ar. The magnetic field has a strong influence on the first group of gases, and a significant but less influence on the second group of gases.

The effect of magnetic field on dielectric gas phase breakdown could be best visualized by comparing the dependence of the energy transfer parameter, W/FM, on pressure with and without magnetic field, depicted in Figure 6.17.

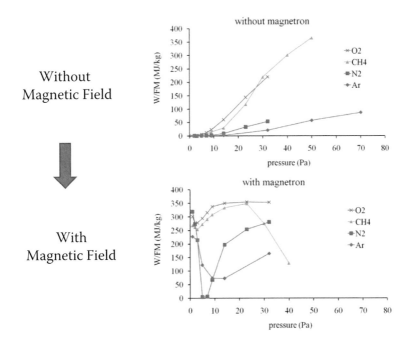

FIGURE 6.17
Effect of magnetic field on gas phase breakdown compared by the dependence of W/FM parameter on system pressure.

The use of W/FM parameter provides a link to gas phase breakdown of RF discharge, in which breakdown voltage and current cannot be measured.

The influence of magnetic field differs significantly depending on the types of gases. However, there are similarity of magnetic field influence in N_2 and Ar, and also in O_2 and CH_4, but those two groups behave differently under the influence of magnetic field. General trends could be described as:

1. Magnetic field shifts the pressure range of operation to the lower pressure domain. Thus, it is clear that the main domain of MLCVD is in the low pressure range, in which stable glow discharge cannot be created by other methods. Accordingly, one of unique advantages of MLCVD is its capability to carry out the plasma polymerization in very low pressure domain. The advantage of low-pressure operation by MLCVD is elucidated in Chapter 9, Section 9.4.

2. At the lowest pressure, very high energy transfer occurs with all gases examined; however, the pressure dependence from the lowest pressure depends on the types of gas mentioned above.

3. With magnetic field, no conspicuous transition point pressure is observed with O_2 and CH_4, but remnant of transition point pressure

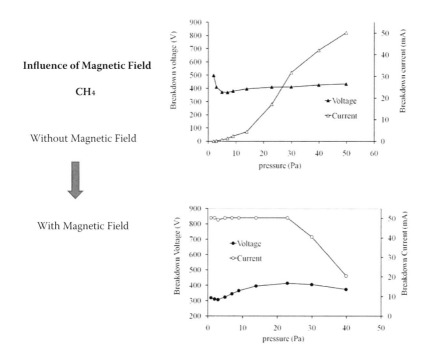

FIGURE 6.18

Effect of magnetic field on gas phase breakdown compared by the dependence of breakdown voltage and breakdown current of methane on system pressure.

is evident with N_2 and Ar, indicating the importance of types of gas in the dielectric breakdown of gas phase.

Figure 6.18 compares the breakdown voltage and current of CH_4 with and without magnetic field. The major influence of magnetic field is seen in (a) huge increase of breakdown current, (b) significant decrease of breakdown voltage, and (c) absence of transition point, which is clearly seen in both breakdown voltage and current, without magnetic field. The differences between the upper figure and the lower figure clearly point out that the dielectric breakdown process with magnetic field is completely different from that without magnetic field; in other words, MLCVD operates in the different domain that could be designated as Domain M in contrast to Domain II and Domain III of gas phase breakdown without magnetic field.

Thus, the superimposed magnetic field changes the phase diagram of the gas phase breakdown for deposition gas CH_4 completely, which means that the gas phase breakdown under the influence of a superimposed magnetic field is a totally different phenomenon from the gas phase breakdown without a magnetic field. DC and AF discharge with magnetron could be operated in the low pressure that cannot be operated otherwise, and the discharge current is much higher than that observed without a magnetic field. These two factors provide unique advantages of MLCVD over other modes of plasma polymerization with respect to the characteristics of nano-film, as discussed in Chapter 7.

The most significant and mysterious fact revealed by examination of the dielectric breakdown of the gas phase under the influence of a magnetic field is the very high current derived by very low applied voltage at a very low-pressure domain. The discharge current observed in the low-pressure domain is at least an order of magnitude greater than what has been observed without a magnetic field using the same reactor, gas, and set of operation parameters. All those surprising findings are not beyond our comprehension of the gas phase breakdown process, if we recognize that the luminous gas phase created is not ionized gas but a phase consisting of highly conductive radiant matters. The big question remains, how could such a high current be created by lower applied voltage? One plausible explanation follows.

6.4 Electrons in Electric Field and in Magnetic Field

Electrons flow through an electrically conducting medium from the cathode to the anode under an applied electric field (i.e., electrons follow the electric field). In gas discharge *without magnetic field*, the current depends on the amount of gas and dissociable gas yield higher conductivity (see Figure 5.33, Chapter 5). The high current results from the capability of the gas phase to move electrons through the gas phase along the electric field lines, which requires a higher number of gas particles per unit volume. The transition point pressure defines the minimum mass concentration in the gas phase which allows the sizable current to develop. Thus, the discharge current increases with the system pressure, and the rate of increase of current is dependent on the degree of dissociation of gas, as discussed in Chapter 5.

Electrons also move from one magnetic pole to the opposite magnetic pole along the magnetic field lines through paramagnetic medium. The motion of electrons in the gas phase under the electric field is linear in the inter-electrode space of two parallel electrodes. The motion of electrons along a magnetic field is spiral along the magnetic field lines, which are curved lines from one pole to the opposite pole projecting in the gas phase.

The interaction of two modes of electron fluxes occurs where two fluxes are orthogonal. That is where the toroidal glow develops. What is not clear is how such a low-voltage/very high–current discharge develops when the electric field is applied in an attempt to initiate magnetron discharge. It is not the case that the normal discharge develops first and then transforms into a toroidal glow. There are some plausible factors to cast light on this puzzle. One is the possibility of changing distribution of free electrons on the cathode surface due to the existing magnetic field. However, efforts to detect possible change of surface conductivity with and without magnetic field turned out to be futile. Another question that arises is are electrons in the magnetic field extending in the gas phase immune to collision with gases?

Before discussing this problem further, it is necessary to recognize some difference in the context of terminologies involved. Apart from the atomic- or molecular-level interpretation of paramagnetic materials and diamagnetic materials, it is necessary to recognize the difference in the phenomenological point of view. In the phenomenological point of view, a paramagnetic material allows the magnetic field of a magnet to penetrate the solid phase of the material in contact, and a diamagnetic material does not allow the magnetic field to penetrate the material phase. Likewise, the magnetic field could be recognized as the special motion of electrons from one magnetic pole to the opposite magnetic pole.

In the gas phase, which is a noncondensed phase, the terminologies dealt in the solid phase do not seem to apply without modification of context based on the phenomenological viewpoint. Hence, paramagnetism and diamagnetism in the context applicable in the solid phase do not apply in the gas phase, because there is no material phase in which atoms or molecules are fixed at specific locations. On the other hand, the magnetic field from a magnet extends to the surrounding gas phase regardless of what kinds of gas exist, even in vacuum, and the magnetic field strength and its distribution can be measured. In this context, all in the gas phase including vacuum are paramagnetic. Although such a phrase as "oxygen is paramagnetic" has been used, it seems to describe the level of electron affinity of oxygen that can be utilized in oxygen-sensing devices. The term *paramagnetic* in the gas phase cannot be used with the same context used for the solid phase of matter.

It is important to recognize that the magnetic field consists of electrons that move from one magnetic pole to the opposite magnetic pole along the magnetic field line in spiral motion around the magnetic field line. It is also very important to realize that there is no mechanism or force in the gas phase that prevents collision of electrons in the magnetic field with gas atoms or molecules. Although the motion of electrons is confined by the magnetic field, the motion of gas molecules is unrestricted. So far as electron-impact reactions with gas molecules are concerned, there is no fundamental distinction between electrons in an electric field and electrons in a magnetic field except that the energy level of electrons and modes of electron movement are different.

Collisions of electrons in magnetic field and gas molecules would occur, but such collisions are gases in random motion colliding with electrons in a circular motion around the magnetic field line. Electrons in the magnetic field are not being accelerated within a magnetic field (i.e., the electron energy in a magnetic field entirely depends on the magnetic field strength). Accordingly, it is anticipated that there is no significant effect on gas per collision. However, one molecule would collide with many electrons, because electrons are circling around the magnetic field, unlike electrons in an electric field, which are rushing to the anode, and the influence of electrons in a magnetic field colliding with gas molecules might reach a level that influences the energy level of gases, depending on the strength of the magnetic field. It was found that if the permanent magnet used to make a magnetron electrode (see Figure 6.2) was weaker than 100 Gauss, no toroidal glow develops, and the presence of the magnetic field has no influence on the gas phase breakdown process.

If the energy level of gas molecules near the cathode surface is raised by repeated collisions with electrons in the magnetic field, the dissociation of such gas molecules by electrons in the electric field, when the electric field is applied to initiate gas phase breakdown, would become much easier, and a large number of energy-transporting species could be created at a given energy level. It seems to be important to remind ourselves that the electrons–molecules interaction in DC discharge is dependent only on the presence of gas molecules in the vicinity of the cathode surface, independent of flow rate, as described in Chapter 8. High currents can be drawn if the conductivity of the luminous gas phase is high. The conductivity, which is given by (I/V), in the breakdown process with a magnetic field is very high, and the conductivity of oxygen, which has higher electronegativity than N_2, is much higher than that of nitrogen in DC discharge without a magnetic field as seen in Figure 5.33 (Chapter 5). Thus, the energy level of gas molecules could be raised closer to the dissociation energy of the gas by collisions with electrons in the magnetic field.

It is important to note that O_2 yields high electric conductivity in the breakdown process with and without a magnetic field. This might be due to electron spins involved in O_2, which makes O_2 a known *paramagnetic* gas molecule, which is in favor of creating electron-transporting species under the condition of the gas phase breakdown by electron-impact dissociation. What we observe in the dielectric gas phase breakdown of O_2 is certainly different from the view of gas phase breakdown of Ar by the ionization principle, and the phenomenon could be understood by reviewing characteristics of the broken-down gas phase which might have been interpreted by different viewpoints (e.g., *paramagnetic* nature of gas, electron negativity, reactivity in luminous gas phase, iN/Out rule, etc.) In any case, the electron-impact reactions of gas molecules with electrons in the magnetic field seem to play a key role in the gas phase breakdown under the influence of a magnetic field.

6.5 Implications of Magnetron Gas Phase Breakdown

6.5.1 Magnetron Chemical Vapor Deposition versus Magnetron Sputtering of Cathode Metal

The electrodes used in magnetron plasma polymerization are essentially the same as the magnetrons used in the magnetron sputtering of the cathode target metal, though magnets are not fixed in the target in the sputtering process. The laboratory magnetron plasma polymerization reactor has been used in the comparative study of magnetron plasma polymerization and magnetron sputter coating using Ar as the gas and the metals with high sputtering yields as the electrode [8]. In other words, the same magnetron reactor could be used for the magnetron sputter coating or for the magnetron plasma polymerization depending on the selected operational conditions of the reactor. However, once the operational conditions were tuned to one process (type of gas, magnetic field strength, discharge power, operation pressure, etc.), the other process usually does not occur simultaneously, because only one gas is used for one type of operation.

Table 6.2 summarizes the main differences of magnetron sputtering and magnetron plasma polymerization. The key factors that separate magnetron plasma polymerization and magnetron sputter coating are the gas used in

TABLE 6.2

Comparison of Direct Current (DC) Plasma Sputter Coating and Magnetron-DC Plasma Polymerization

	DC Plasma Sputter Coating	Magnetron-DC Plasma Polymerization
Controlling parameter	Current density	Current density
Creation of depositing species	Impact of low-energy electrons in the linear motion toward anode	Impact of low-energy electrons circling along the magnetic field line
Location of activation	Toroidal glow (on) the cathode surface	The toroidal dissociation glow is in the gas phase near the surface of electrodes
Confinement of glow	Yes	Yes—more confined
Yield of deposition on substrate in gas phase	High	
	High; depending on the distance from the toroidal glow	
Deposition on the cathode surface	Low	
	Low; a part of circular deposition on the cathode surface has very little to no deposition	
Wall contamination by plasma deposition	Little/moderate	Very little
Change of plasma deposition behavior with reaction time	Little with moving location of magnet	Negligible
Continuous long-time operation	Difficult	Possible
Scale-up of process	Bound to be batch process	Possible (?)

the processes, the metal used as the electrode, and the level of energy input for the discharge.

6.5.2 Magnetron Discharge Sputtering

The magnetic field does not change the fundamental sequence of the gas phase breakdown and subsequent electron-impact processes, but it changes the location of the main electron flux from the cathode surface and the efficiency of electron-impact processes. The comparative study of magnetron sputter deposition and magnetron plasma polymerization (MLCVD) by the same reactor indicated that the sputtering by the magnetron discharge does not occur by the momentum exchange mechanism.

The major experimental facts that negate the momentum exchange mechanism were found in the effect of magnetic field strength on the discharge characteristics and also on the sputter deposition rate. The sputter-coating rate increases with increasing magnetic field strength as shown in Figure 6.19. However, the magnetic field strength also changes the glow discharge characteristics of Ar and of CH_4 in the same manner as depicted in Figures 6.20 and 6.21, respectively. The important fact is that the superimposed magnetic field changes the high voltage–low current discharge in nonmagnetron discharge to the low voltage–high current discharge in the magnetron discharge, both for Ar and methane. Furthermore, the higher is the magnetic field strength, the lower is the voltage and the higher is the discharge current. Those findings are what we anticipate from the domain [M] described in Section 6.3. A serious question that arises is as follows: How can lowering the discharge voltage increase the momentum of ions? Of course, it is impossible, because higher momentum of ions requires higher accelerating voltage. Hence, it is clear that

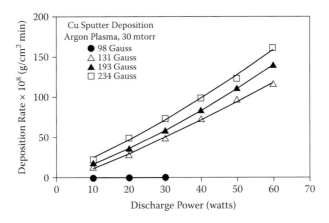

FIGURE 6.19
Rate of sputter deposition of copper as a function of discharge power in argon plasmas (30 mtorr) with various magnetic fields.

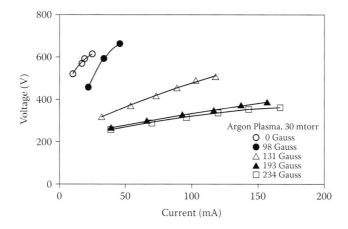

FIGURE 6.20
V-I curves for the argon magnetron glow discharge plasma at a pressure of 30 mtorr.

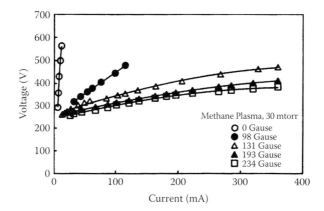

FIGURE 6.21
V-I curves for the methane magnetron glow discharge plasma at a pressure of 30 mtorr.

the momentum exchange principle of impinging ions to dislodge metallic atoms does not occur, at least under the conditions used in these comparative studies.

The only positive correlation found is between the discharge current and the sputtering rate (i.e., the sputtering rate increases with the discharge current, not discharge voltage). Because the avalanche of the primary electrons from the cathode surface occurs at the toroidal glow surface of the cathode, the bombardment of the accelerated ions onto the toroidal glow cathode surface is an unreasonable concept.

The electrode surface corresponding to the toroidal glow reveals that the middle of the toroidal glow surface often has dark powdery deposition, and there are shiny surfaces without polymer deposition on both sides of the dark deposition in magnetron plasma polymerization. The dark powdery deposition does not attach to the surface and drops off from the surface as the amount of deposition reaches the threshold value. Depending on the detail of operational conditions and the length of operation, the toroidal glow surface does not show deposition.

It is important to note that the location of a no-deposition surface in magnetron plasma polymerization coincides with the location of highest sputtering (ablation) on the electrode surface in a magnetron sputter deposition process when the same reactor system is used for sputtering processes. Therefore, it seems quite logical to consider that the high flux of electrons from the surface is responsible for the ablation of metal from the cathode surface in the sputtering operation and also for the lack of deposition of plasma polymer in the magnetron plasma polymerization.

It had been considered that sputtering occurs as a plasma-enhanced thermal process (e.g., thermal evaporation) during the approximate period of 1935 to 1955. Since then, the concept of the momentum exchange principle has been widely accepted [9,10]. Both mechanisms assume the bombardment of accelerated ions on the cathode surface. However, there is no experimental evidence presented in the literature that accelerated ions strike the toroidal glow cathode surface under the conditions being discussed. There are experimental data that show that the required momentum of impinging ions to cause the sputtering of the cathode metal at the toroidal glow magnetron surface decreases as the sputtering rate increases with the magnetic field strength. The comparative studies of the magnetron sputtering and the magnetron plasma polymerization support the mechanism that could be described as "the primary electron avalanche induced ablation or sublimation of the cathode metal," which is more on the side of the earlier interpretation of the sputtering so far as the mechanism of the ablation of metal atoms is concerned. Figure 6.22 depicts the difference between MLCVD and magnetron sputtering processes [12].

6.6 Magnetic Field Initiation of Luminous Gas Phase

6.6.1 Collisions of Gas Molecules with Electrons in Magnetic Field

The data shown above clearly indicate that the gas phase breakdown process with magnetic field is totally different from that without magnetic field. The most striking feature is that high current flow through the broken-down gas phase, which contains a very low quantity of gas, and the breakdown voltage to create such a large current are surprisingly low. In domain II of the broken-down gas phase without magnetic field, high voltage is necessary to create very low current (i.e., *high voltage/low current*). However, the fact that

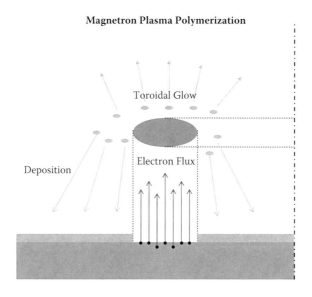

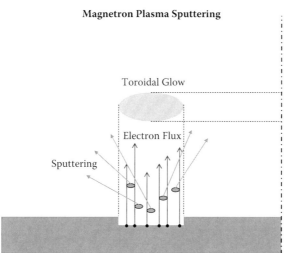

FIGURE 6.22 (*See color insert.*)
Magnetron plasma polymerization (top) and magnetron plasma sputtering (bottom).

the influence of the magnetic field is most pronounced with dissociable gas methane, and molecular gas with high electron affinity, oxygen, means that the same electron-impact mass-actions, which cause the gas phase breakdown without magnetic field, are responsible for the gas phase breakdown under the influence of the magnetic field.

It seems reasonable to consider that electrons bound to the magnetic field are not immune to the collision with gas particles as discussed in the beginning of this section. Furthermore, if the energy transfer from electrons bound to the magnetic field to gas molecules occurs as explained for unusually high current observed with magnetron discharge, the possibility of the magnetic field causing the onset of luminous gas phase seems to be realistic, if the magnetic field becomes strong enough so that electrons in the magnetic field carry enough energy to cumulatively dissociate gases. If the energy with electrons in a magnetic field is high enough, this leads to the possibility of a larger number of electrons leading to the electron-impact mass actions, even in the absence of an electric field, and creates larger numbers of dissociated species that act as the electron energy-transferring medium (i.e., magnetic field inception of the luminous gas phase). It is important to keep in mind that the overall impact of an electron colliding with a gas molecule is highly dependent on the ability of the gas to dissociate and the electronegativity of atoms in the molecule.

6.6.2 Potential Mechanism for Inception of Aurora Borealis

The recent National Aeronautics and Space Administration's (NASA) THEMIS Project revealed that the appearance of the aurora borealis (northern lights) is caused by an electromagnetic field burst or magnetic slingshot from the earth. The critically important point, from the concept of dielectric breakdown of the gas phase, is that aurora inception occurs first against the prediction that the aurora inception should follow the return of high energy to the earth due to the fusing of magnetic fields caused by a magnetic field burst (i.e., the aurora inception is expected to be the last step of the magnetic field burst from the earth). The sequence found (i.e., the aurora first) suggests that the magnetic field triggers the onset of aurora. The conventional mind-set seems to seek a possible source of high energy to cause gas phase ionization of gases in order to explain the mechanism for the inception of aurora, while the creation of a luminous gas phase in low pressure under the influence of a magnetic field occurs with a very low level of energy, as shown in this chapter.

As discussed earlier in this section, applying the conventional wisdom obtained with mono-atomic gas to the breakdown process of molecular gases is irrelevant, and so seems to be the application of high-energy ionization of gases in low pressure to aurora inception. There seems to be subtle but important distinction between the cause of magnetic field burst from the earth and the inception of aurora. Although the source of energy to create aurora has been clarified, the mechanisms of how the energy is transferred to gas molecules and creates a luminous gas phase at very low pressure seem to be left unsolved or unaddressed.

The effects of magnetic field on the breakdown process seem to cast light on the mechanism of how the aurora could be induced by the magnetic field burst. Accordingly, by employing an electro-magnetron, rather than permanent magnets, the magnetic field induction of the luminous gas phase (plasma) could be investigated. A preliminary effort seems to suggest very promising indications, although much more fundamental investigations along this line of thought are needed in the future. The highlights of promising indications are briefly expressed as follows.

When a voltage that is below the breakdown voltage of gas (N_2) without a magnetic field is applied at a low pressure below the transition point pressure, no discharge occurs as the phase diagram of the gas phase breakdown tells us; the applied voltage and pressure used are in domain [I] where no gas phase breakdown occurs. After confirming that selected conditions are in domain [I], the electric magnet (solenoid) is activated, and the current is increased; the magnetic field is increased. As a certain current is reached, glow discharge occurs with strong luminosity and high current, which is what we anticipate happening according to the mechanism suggested previously. However, this trend decreases as the preset pressure increases toward the transition point pressure, and at near the transition point pressure and above, the effect of the magnetic field becomes negative, which seems to follow the trend observed with the gas phase breakdown of N_2 described in Section 6.3. Due to the inability to create a stronger magnetic field by a solenoid, experiments with pure magnetic field inception of glow discharge could not be completed; however, it seems that the magnetic field inception of the luminous gas phase is a very likely event.

If we could initiate the luminous gas phase by the magnetic field, without the electric field applied, the characteristics of the luminous gas phase would be significantly different from those of glow created by DC discharge with magnetic field. With an applied electric field, electrons that caused the gas phase breakdown move straight to the anode, and if the applied DC voltage is shut down, the discharge extinguishes immediately. If the electric field was created by some unintentional electric field, such as in the case of lightning, the discharge would extinguish as soon as the electric potential is nullified by the electron flux.

The luminous gas phase created by the magnetic field alone would remain as long as the magnetic field that caused the inception of luminous gas phase exists, maintaining the luminosity by the principle of reexcitation of the photon-emitting excited species by low-energy electrons, as described in Section 6.2.

These anticipated characteristics of a magnetic field–induced luminous gas phase could explain some unique and mysterious behavior of the aurora:

1. Aurora does not appear all the time, but appears only when the magnetic field burst from the earth occurs, as NASA experiments using four satellites unequivocally proved (i.e., aurora occurs only with an additional magnetic field [magnetic field burst]).

2. Aurora remains for a long time (e.g., a few hours), while an electrical potential difference–triggered discharge, such as lightning, remains

only for a few seconds. As soon as the electric potential gradient that caused the lightning is nullified, the lightning ceases. If aurora was initiated by an electric field created by any other cause, the duration of aurora should be as short as lightning.

3. Most importantly, the entire aurora, the aurora luminous gas phase, is not uniform, indicating that the formation of luminous gas occurs at wide areas, not at any specific location, unlike dissociation glow on the cathode surface.

4. Aurora is not stationary and moves around, because neutral photon-emitting species move randomly as gases (no directional move-ment), and there are layers of magnetic field lines in the atmosphere that can reexcite photon-emitting neutrals that have just lost energy by emitting photons. The strength of the magnetic field decreases as the distance from the earth increases, and the magnetic field cannot create the luminous gas phase or sustain the luminous gas already created, which sets the boundary of aurora.

5. It is highly likely that the magnetic field strength fluctuates during the period of magnetic field burst, which will change the location of excitation as well as the boundary of aurora. In combination with aspects described above (4), it causes the most amazing and unpre-dictable movement of the overall luminous gas phase, aurora.

There seem to be many strikingly parallel phenomenological aspects between the aurora inception mechanism described in this section and the gas phase breakdown under the influence of a magnetic field. Further inves-tigation of the reaction mechanism of MLCVD might lead to further eluci-dation of the aurora inception mechanism to such an extent that the term *MLCVD* should be changed to *aurora CVD*.

References

1. Tao, W. H., M. A. Prelas, and H. K. Yasuda, *Journal of Vacuum Science and Technology. A, Vacuum, Surfaces, and Films*, 14(4), 2113, 1996.
2. Tao, W. H. and H. K. Yasuda, *Plasma Chemistry and Plasma Processing*, 22(2), 297, 2002.
3. Yasuda, H. and Q. Yu, *Plasma Chemistry and Plasma Processing*, 24, 325, 2004.
4. Yasuda, H., *Plasma Processes and Polymers*, 2, 293–304, 2005.
5. Yasuda, H., *Luminous Chemical Vapor Deposition and Interface Engineering*, CRC Press, Boca Raton, FL, 2004.
6. Yasuda, H., *Plasma Processes and Polymers*, 4, 347–359, 2007.
7. Ledernez, L., F. Olcaytug, G. A. Urban, and H. K. Yasuda, *Plasma Chemistry and Plasma Processing*, 6, 659–667, 2007.

8. Cho, D. L., Y. S. Yeh, and H. Yasuda, *Journal of Vacuum Science and Technology. A, Vacuum, Surfaces, and Films*, 7, 2960, 1989.

9. Vossen, J. L. and J. J. Cuomo, *Thin Film Processes*, J. L. Vossen and W. Kern, Eds., Academic Press, New York, 1978.

10. Wasa, K. and S. Hayakawa, *Handbook of Sputter Deposition Technology*, Noyes, Westwood, NJ, 1992.

11. Yasuda, H. et al., Data to be published, 2010.

12. Yasuda, H., *Plasma Processes and Polymers*, 5, 215–227, 2008.

7

Polymer Formation Mechanism in Luminous Gas

7.1 Free-Radical Polymerization and Free-Radical Polymer Formation in Luminous Gas Phase

Nearly all polymers formed by plasma contain large amounts of free radicals detectable by electron spin resonance (ESR), and there is no doubt that polymer formation proceeds with free radicals. However, the presence of a large number of free radicals in deposition is in contradiction with the conventional free-radical polymerization mechanism (i.e., polymers formed by conventional free-radical addition-polymerization do not contain sizable free radicals, because the recombination of two growing molecules with a free radical at the growing chain-end is the termination process of free-radical polymerization). Furthermore, organic molecules that are not monomers of free-radical polymerization (e.g., saturated vinyl monomers) form polymers just as easily as corresponding vinyl monomers in the luminous gas phase (plasma). In plasma polymerization, reactive species with free radicals are created by the electron-impact dissociation of molecules, and many gases used in plasma polymerization (e.g., methane, benzene, etc.) do not have functional groups for free-radical addition-polymerization. In other words, the evidence that plasma-polymerized polymers contain large amounts of free radicals is proof that free-radical plasma polymerization is not conventional chain growth free-radical polymerization.

In conventional free-radical polymerization, the monomer must have a chain-carrying functional group (e.g., double bond or triple bond), and a small concentration of a free radical-forming initiator (e.g., peroxide) is added to a reaction mixture. The concentration of the initiator determines the molecular weight of resultant polymers—that is, the greater the concentration of initiator, the shorter is the kinetic chain length, and polymers with a lower degree of polymerization are formed; the initiation and the termination are coupled. In plasma polymerization, no initiating chemical is added in the polymerization system, and no particular functional group for chain-growth polymerization is necessary. A vinyl monomer and its saturated monomer polymerize by a similar polymerization rate in plasma polymerization, while saturated monomers do not polymerize in free-radical polymerization. The chemically reactive species to form polymers are created by the electron-impact dissociation of molecules.

The electron-impact dissociation of molecules (breaking of any covalent bond including π bond) yields free radicals, but those free radicals mainly recombine each other, which is the termination step of conventional free-radical polymerization. Because so many free radicals are formed within a short time span, the kinetic chain growth is zero (i.e., no polymerization by conventional free-radical addition polymerization mechanism). This situation is analogous to the case that too much initiator is added to the free-radical polymerization system (i.e., no polymer formation because the growing chains recombine with abundantly available growing molecules with free-radical end before enough addition of monomer molecules occurs, yielding formation of many low molecular weight oligomers).

The polymerizable functional groups, for the conventional free-radical polymerization, present in gas molecules only influence the mode of electron-impact dissociation, but its influence is relatively small, as seen in Table 7.1, which compares the deposition rate of the monomer of conventional free-radical polymerization and the corresponding saturated (without double bond) molecules in the luminous gas phase [1]. It should be kept in mind that in conventional free-radical polymerization, molecules shown on the right column have a polymerization rate of zero—no polymerization. Table 7.2 lists types of molecules and their reaction characteristics in plasma polymerization.

7.2 Repeating Step Growth Polymerization (RSGP) Mechanism

Plasma polymerization in the luminous gas phase proceeds via repeated recombination of free radicals and re-excitation, which means the recombined species are subjected to reactivation (formation of free radicals) to form free radicals on the once-recombined molecules. The dissociation of molecules, including the products of recombination of free radicals and molecular species with a free radical, occurs by the electron-impact dissociation in the luminous gas phase. The complex polymer formation mechanisms are expressed by the repeating step growth polymerization or rapid step growth polymerization (RSGP) mechanism, because the molecular weight increase occurs stepwise but in rapid repeating mode [2]. The overall mechanisms of polymer formation and deposition are schematically depicted in Figure 7.1 in a revised format that includes deposition steps.

TABLE 7.1

Comparison of Deposition Rates for Vinyl Monomers and Corresponding
Saturated Compounds

Vinyl Compounds	k^a	Saturated Vinyl Compounds	k^a
N⟩—CH=CH₂	7.59	N⟩—CH₂—CH₃	4.72
⬡—C(CH₃)(CH₂)	5.33	⬡—CH(CH₃)(CH₃)	4.05
⬡—CH=CH₂	5.65	⬡—CH₂—CH₃	4.52
H₃C—⟨N⟩—CH=CH₂	7.65	H₃C—⟨N⟩—CH₂—CH₃	7.38
N—CH=CH₂ (cyclic, O)	7.55	N—CH₂—CH₃ (cyclic, O)	3.76
$H_2C=CH-C\equiv N$	5.71	$H_3C-CH_2-C\equiv N$	4.49
$H_2C=C(Cl)(Cl)$	5.47	$H_3C-CH(Cl)(Cl)$	2.98
$H_2C=CH-CH_2-NH_2$	2.86	$H_3C-CH_2-CH_2-NH_2$	2.52

Note: Parameter k is expressed in units of cm^{-2} × 10^4; $r = kF_w$, where r is the rate of
polymer deposition (g/cm²•min), and F_w is the weight-based monomer flow
rate (g/min).

Electron impact on the deposition molecular gas, M, creates free-radical
bearing species $M\bullet$. If a gas molecule contains a di-radical forming chemical structures (e.g., double bond, triple bond, cyclic structure, etc.), di-radical
bearing species $\bullet M\bullet$ are formed. The overall polymer formation steps consist
of three repeating cycles: mono-radical cycle-1($M\bullet + \bullet M$ cycle), mono-radical
cycle-2 ($M\bullet + \bullet M\bullet$ cycle), and di-radical cycle ($\bullet M\bullet + \bullet M\bullet$ cycle). Without di-
radical forming chemical structures, the overall polymer formation mecha-
nism reduces to $M\bullet + \bullet M$ cycle only. This is one of the key reasons the use of
simple organic molecules in magneto-luminous chemical vapor deposition
(MLCVD) (e.g., methane and trimethylsilane) is preferred, making the processing simpler and reducing the amount of trapped residual free radicals in
the product.

TABLE 7.2

Classification of Types of Monomers

Type	Chemical Structure	Luminous Chemical Vapor Deposition (LCVD) Reaction Characteristics	Characteristic Features of LCVD Deposits
I	Aromatic Hetero-aromatic Triple bond	Polymerize readily with little hydrogen production and low photon emission	Polymer contains high level of trapped free radicals and unsaturation
II	Double bond Cyclic structure	Polymerize with moderate hydrogen production and moderate photon emission	Moderate level of trapped free radicals and unsaturation
III	Saturated hydrocarbons	Polymerize with high yield of hydrogen and high level of photon emission	Low concentration of trapped free radicals and unsaturation
IV	Oxygen (in aliphatic structure)	Low deposition rate, high rate of byproduct gases, high level of photon emission (poisoning effect)	Very sensitive to the energy input level of LCVD; often the oxygen-containing groups are absent in polymers
V	Oxygen (in cyclic structure including epoxides)	Oxygen atom is preferentially removed yielding di-radical for polymerization	Oxygen is nearly absent in polymers
VI	Oxygen (in Si-containing structure)	Polymerize with low yield of hydrogen	Oxygen remains in deposition
VII	Si containing	Si-containing species deposit faster than C-based species	Stable polymer depositions; deposition (on reactor) cannot be etched by O_2 plasma; addition of O_2 yields varieties of hydrophilicity of deposition
VIII	F containing	Behave completely different from the rest of monomers, and reproducibility is poor	Types of F in deposition depend on many operation factors and location in a reactor

Important features shown in Figure 7.1 are as follows:

1. The formation of free radicals and recombination of free radicals occurs independently (i.e., these two kinds of reactions are not coupled). Under a high free-radical formation rate, the recombination cannot catch up with the formation rate, which leads to a large number of unreacted free radicals left in the deposition (plasma polymer) and leaves unreacted free radicals in plasma polymerization products.

2. The initiation reaction that is the first step of the chemical reaction, without which the whole process does not proceed, and the deposition steps that bring down species in the polymer formation

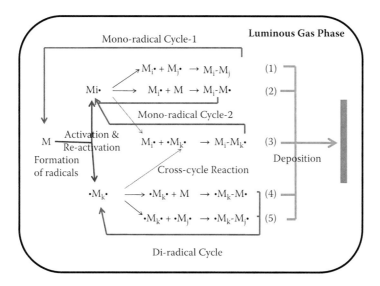

FIGURE 7.1 (*See color insert.*)
Polymer formation mechanism in luminous gas phase: RSGP mechanism.

steps in the luminous gas phase to the substrate surface are also not coupled. Any species described in reaction (1) through reaction (5) could deposit on the substrate surface and on any other surface that contacts the gas phase.

Because of these two features that are built in the plasma polymer formation mechanisms, plasma polymerization, in general, is highly system dependent. Accordingly, a generic plasma polymerization of styrene or any other molecule cannot be defined. In other words, various kinds of plasma polymer could be prepared, but not a single product from a monomer (e.g., styrene) by plasma polymerization.

In this chapter, we are dealing with how the monomer gas interacts with the luminous gas phase, which was created by the dielectric breakdown process described in Chapter 5. Monomer gas first interacts with the energetic luminous gas phase, which occurs where the gas introduced into the reactor meets with an already established luminous gas phase. However, the major events take place at the most intense glow, which could be described as the energy-input sheath of the luminous gas phase where the major energy transfer from electrons to deposition gas molecules occurs. This assessment is based on the fact that the glow characteristics do not change once the steady-state luminous gas phase is established a few seconds after the onset of glow discharge and the fact that the kinetic study of direct current (DC) cathodic polymerization indicates the monomer gas is drawn into the cathodic dissociation glow due to a decrease in partial pressure of the monomer gas, which is described in Chapter 8. This assessment is particularly true with

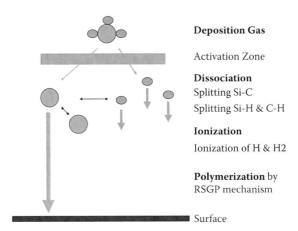

Deposition Gas

Activation Zone

Dissociation
Splitting Si-C
Splitting Si-H & C-H

Ionization
Ionization of H & H2

Polymerization by
RSGP mechanism

Surface

FIGURE 7.2
Sequence of reactions involved in the formation of polymeric deposition in the luminous gas phase.

MLCVD, because it has been shown that reproducible products have been produced in 1-month-long continuous operations in many reactors with the same design.

Figure 7.2 schematically describes the sequence of chemical reactions that occurs in this process. The deposition gas shown in the simplified model is trimethylsilane (TMS), $HSi(CH_3)_3$, without showing H atoms in the molecule. The activation zone is the location where the dissociation of molecules occurs. Dissociation reactions include the splitting of an H atom from Si-H and CH_3. Although splitting H reactions, mainly of C-H with TMS, are not shown in the figure, those are main dissociation reactions as seen in the increase of system pressure shown in Figures 5.8 and 5.9 (Chapter 5). The primary reaction is the dissociation of the TMS molecule. The main ionization is that of H, as shown in Chapter 5, which is the ionization of the dissociation product (i.e., the ionization is the secondary reaction in plasma polymerization). Various species created by the dissociation of TMS molecules yield polymeric deposits via the RSGP polymerization mechanism depicted in Figure 7.1.

The electron-impact dissociation of organic molecules can proceed in various modes depending on the chemical structure of molecules and the energy input level of the operation. In saturated hydrocarbons, the main dissociation is the detachment of H by breaking the C-H bond. Hydrogen atoms created by the dissociation predominantly recombine to form hydrogen molecules, H_2, which is not shown in Figure 7.1, because H_2 does not contribute to the increase of molecular weight of gaseous species. A covalent σ bond consists of two electrons paired. When a monomer molecule, M, undergoes electron-impact dissociation, a monomer free radical $M\bullet$ is formed, of which subsequent chemical reactions are depicted in the upper half of the figure with the designation of $M_i\bullet$ and $M_j\bullet$. The suffixes i and j indicate that the molecular size could be different depending on how many times the recombination and reexcitation had occurred, but both M_i and M_j are not the original monomer M, and their molecular weight is different (generally greater) than that of M. Suffixes i and j simply indicate M_i, M_j, and M are not the same, and they are used without specific numerical values.

Reaction (1) depicts the recombination of two free radicals yielding a molecule without a free radical site. With saturated hydrocarbons, reaction (1) with reexcitation is the main polymer formation path. In the presence of double or triple bonds or cyclic structure in the deposition gas molecule, reactions (2) through (5) come in to the RSGP mechanism. Any species shown on the figure, including the reaction product of reaction (1), could deposit on the substrate surface or repeat the corresponding cycle of reexcitation and recombination of free radicals. Every repetition of reaction cycle increases the sites for branching or cross-linking.

Reaction (2) depicts the addition of the free radical caused by the hydrogen abstraction reacting with a molecule with a functional group that preferentially reacts with a free radical (e.g., a double bond, a triple bond, etc.). The reaction product is a free radical with larger molecular weight (mono-radical cycle-2). Reaction (2), however, does not lead to free-radical addition polymerization as described before.

The electron-impact reaction of molecules with a double bond, a triple bond, or cyclic (ring) structure preferentially occurs in the opening of a double bond, a triple bond, or a cyclic structure, which yields two unpaired electrons (free radicals) within a molecule, which is depicted in the lower half of Figure 7.1—reaction (4) and reaction (5). The reaction products of those reactions are essentially identical to the starting chemical moiety (i.e., molecules with two free-radical sites). Reaction (4) and reaction (5) constitute the di-radical cycle, without reactivation. This cycle serves as the reservoir of di-radicals in the luminous gas phase. If the starting monomer molecule has little hydrogen (e.g., acetylene, benzene), the polymer-forming reaction proceeds rapidly with the recombination of bi-radicals. It is important to note that benzene, which is a solvent used in free-radical polymerization (solvent does not polymerize) forms plasma polymer nearly as fast as acetylene because of this reaction mechanism.

If both mono-free radicals and di-free radicals exist in the luminous gas phase, the reaction between mono-radical and di-radical is inevitable, which is depicted in reaction (3) (cross-cycle reaction). The cross-cycle reaction converts the di-radicals in the di-radical cycles (reaction (4) and reaction (5)) to mono-radicals, which reduces the number of di-radicals in the luminous gas phase. The role of the cross-cycles reaction became clear when pulsed discharge was employed; the number of trapped free radicals in the deposition of a double bond containing monomer increased dramatically [3], as shown in Table 7.2 [4]. During the off period of pulsed discharge, the production of mono-radicals and di-radicals from the monomer ceased, and reaction (3) also ceased, but the di-radical cycle—reactions (4) and (5)—and reaction (2), the mono-radical addition reaction, if the monomer has a structure with a multiple bond, continue to proceed, yielding a larger number of free radicals on the surface of the substrate than in the corresponding continuous discharge, in which a considerable portion of di-radicals are converted to mono-radicals by the cross-cycle reaction (3).

The major difference in the polymerization mechanisms in conventional free-radical polymerization and free-radical plasma polymerization could be visualized by the following schematic expressions of the growth mechanisms. The growth mechanism of the conventional free radical is the repeating additions of a free radical, which is created by the action of an initiator molecule, to a π bond of the monomer molecule, $n(\bullet + M)$, where n is the number of repeating, M is the monomer molecule, and $\bullet$ represents an unpaired electron (free radical). One free-radical site reacts with n number of monomer molecules to create a polymeric free-radical molecule with the degree of polymerization n (i.e., $\bullet M_n$). On the termination reaction of free-radical polymerization, two polymeric free radicals $\bullet M_n$ and $\bullet M_m$ recombine to form a polymer $(M_n - M_m)$.

In contrast to this situation, the growth mechanism of free-radical plasma polymerization could be given by a similar notation as the repeating recombination of free radicals formed by the dissociation of the monomer by the electron-impact dissociation, $(\bullet + \bullet = M_i - M_j)^n$, where n is the number of times the whole sequence of elementary steps of plasma polymerization is repeated (i.e., electron-impact dissociation and recombination of free radicals). The symbol $(\bullet)$ represents the free-radical site of the reactive species. The recombination of free radicals creates a σ-bond (no free radical), and the molecule formed by one step is subjected to the electron-impact dissociation process repeatedly (n times).

In the dissociation processes, π-bond in a double or triple bond has the lowest bond energy; however, the ease of opening a π-bond by the impact of an electron depends on the surrounding chemical moieties. For instance, the electron-impact dissociation of a molecule of benzene yields 3 $(\bullet C = C \bullet)$, which is the identical structure $(\bullet C = C \bullet)$ formed by the opening of a π-bond of the triple bond of acetylene (i.e., benzene behaves as if it is three acetylene in the luminous gas phase). Consequently, the reduced plasma polymerization rate per mass, R_m/FM or R_m/M_v, where R_m is the mass deposition rate, F is the volume flow rate, M is the molecular weight, and M_v is total mass in the luminous gas volume, which is nearly identical to that for acetylene, while benzene is not a monomer of conventional free-radical polymerization. The electronegativity of atoms and bonds dictates the mode of dissociation reactions, of which details are described in Section 5.10, dealing with electronegativity.

The formation of free-radical–containing molecular species occurs by electron-impact dissociation of molecules, breaking chemical bonds, which continues to occur as long as the applied electric field is maintained and a steady flow of deposition gas is maintained. Hydrogen abstraction from monomer molecules plays a significant role in creating free radicals and, hence, in the plasma polymerization rate of hydrocarbon monomers. Examination of a closed-system gas phase after a known amount of a hydrocarbon was subjected to radio frequency (RF) glow discharge revealed the importance of hydrogen abstraction in plasma polymerization [5]. According to the results, nearly all hydrocarbons were converted to polymers, with the yield varying from 85% to more than 99% in a relatively short time under the conditions used, and the gas phase after the polymerization (excluding unreacted organic vapor, which is 0% to 15% of the monomer depending on the polymer yield) consisted mainly of hydrogen.

The hydrogen production expressed as the hydrogen yield per monomer molecule (number of hydrogen molecules evolved when a monomer participates in polymer formation) increases with the increasing number of hydrogen atoms in a hydrocarbon, as depicted in Figure 7.3. In order to distinguish the role of double bond, triple bond, cyclic structure, and aromatic structure, the hydrogen yield is plotted against the structure parameter, which is given by the ratio of (number of hydrogen atoms in a molecule)/(number of structures), in Figure 7.3. For instance, in the case of cyclohexene, the total number of hydrogen, 10, is divided by two structures (i.e., one cyclic structure and one double bond).

There are clear separations of curves depending on the types of monomer structure, which indicate the contribution of a chemical structure in a molecule. There is also a strikingly regular dependence of the hydrogen yield on the number of hydrogen atoms in a molecule (within a group). This smooth and regular dependence strongly indicates that every C-H bond in organic molecules has an equal probability for hydrogen detachment and participant reaction (1) described by Figure 7.1.

The recombination of free radicals requires the collision of larger species than electrons and molecules. Consequently, more free radicals are formed than are dissipated by the recombination. This difference increases with energy input per mass of gas. Thus, the properties of plasma polymer depend on operational parameters of plasma polymerization, while properties of a conventional polymer are largely fixed by the molecular structure of the monomer. Based on data presented above, the types of monomers to be used in plasma polymerization could be classified as shown in Table 7.3. Number represents the percent change due to pulsing from the correspond-

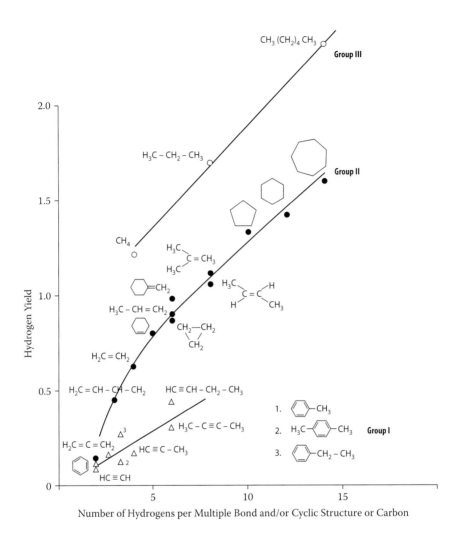

FIGURE 7.3
Number of hydrogen molecules evolved per molecule of starting material when hydrocarbon polymerizes (hydrogen yield) as a function of chemical structure.

ing value for the continuous plasma. $\delta = p_g/p_m$, where p_g is the pressure of the glow discharge, and p_m is the pressure of the monomer flow before discharge. Glass free radical is due to ultraviolet (UV) radiation during plasma polymerization. Cos θ (H_2O) is the value of cosine of the contact angle of water; a negative value means that the plasma polymer becomes more hydrophobic, and a positive value indicates that plasma polymer becomes more hydrophilic by pulsed glow discharge.

The tricyclic RSGP mechanism shown in Figure 7.1 has an important implication for the interpretation of diagnostic data of the luminous gas phase:

TABLE 7.3

Effects of Pulsed Radio Frequency (RF) Discharge on Repeating Step Growth Polymerization (RSGP) Mechanism with Various Monomers

Monomer	Type of Reaction	Cycle in RSGP	Pressure Change	Deposition Rate	Free-Radicals	Glass Radicals	Cos θ (H$_2$O)
Tetramethyl disiloxane	Det. H Det. CH$_3$	1 1	−9	−47	−90	0	−0.28
Propionic acid	Det. H Det. COOH	1 1 and Pois.	−18	−110	0	−81	0.53
Acrylic acid	Det. H Det. COOH Open C=C	1 1 and Pois. 2	−29	120	140	−100	0.49
Cyclohexane	Det. H Rg. Open.	1 2	21	−90	−100	−100	−0.09
Ethylene	Det. H Open C=C	1 2	19	2	970	−79	−0.09
Acetylene	(Det. H) Open triple bond	1 2	200	−23	81	0	−0.01
Benzene	Det. H Ar-Rg. Open.	1 2	150	−8	−50	0	−0.76
Styrene	Det. H Open C=C Ar-Rg. Open.	1 2 2	87	−16	−86	0	−0.09
Vinyl acetate	Det. H Det. OCOMe Open C=C	1 1 2	−5	−48	−21	−71	−0.03
Ethylene oxide	Det. H Rg. Open.	1 2 and Pois.	−5	−7	−33	−76	−0.09
Tetrafluoro ethylene	Det. F Open C=C	1 and Abl. 2	−11	106	−35	−84	−0.01
Hexafluoro-benzene	Det. F Ar-Rg. Open.	1 and Abl. 2	50	−22	−27	0	−0.05
Vinyl fluoride	Det. H Det. F Open C=C	1 1 and Abl. 2	47	−45	480	−100	−0.07
Vinylidene fluoride	Det. H Det. F Open C=C	1 1 and Abl. 2	−6	−47	100	−64	−0.36

Note: Det., detachment; Pois., poisoning; Abl., ablation; Rg., ring; Ar-Rg, aromatic ring.

1. Any species identified in the plasma phase are intermediate species of plasma polymerization but not precursors of "black box" plasma polymerization.
2. Species identified in the deposition plasma could be the species that will not be involved in the polymer-forming reactions (e.g., hydrogen represents the dissociation reaction necessary to create

polymerizable species, but it will not be involved in polymer-forming chemical reactions).

3. Analysis of species in the gas phase of deposition plasma, without understanding the reaction mechanisms to confirm or the hypothetical reaction mechanisms to test, is not warranted, and interpretation could be misleading.

7.3 Competitive Ablation and Polymerization (CAP) Principle

What happens in a low-pressure plasma process cannot be determined in an *a priori* manner based only on the nature of the gas used or on the objective of the process (e.g., plasma etching or plasma deposition), because plasma deposition and plasma ablation occur simultaneously in a competitive manner. The plasma sensitivity series of elements in molecules, both in the luminous gas phase and the solids, which contacts the luminous gas, determines the balance between ablation and polymerization. Once gases evolve from a highly plasma ablation-sensitive substrate, the plasma polymerization system cannot distinguish between monomer and ablated gas, which means plasma polymerization is highly dependent on the whole system, not just gas used, or a particular purpose of deposition. An aimed plasma polymerization could turn out to be plasma ablation, and conversely, plasma etching turned out to be plasma deposition of etching gas when an inadvertently introduced gas impurity plays a key role in shifting the balance between deposition and ablation.

Plasma sensitivity refers to the fragmentation tendency of gaseous materials and of surfaces that come into contact with the luminous gas, which contains various energetic species including ions, electrons, excited species, meta-stables, photons, and chemically reactive species. Ionizing radiation with energetic species such as electron beams, ion beams, and x-radiation, causes much more severe (penetrating) fragmentation than exposure to the luminous gas (low-pressure plasma). Plasma sensitivity refers to the latter case rather than to sensitivity to ionizing (high-energy) radiation.

Plasma sensitivity series refers to the order of element sensitivity to plasma, in a manner similar to the expression of ionization of metals in solution by the galvanic series. There is no clear-cut plasma sensitivity series established today. However, there are some trends that seem to be closely related to the plasma sensitivity series. These are trends found in the order of weight loss rates when polymeric materials are exposed to plasmas [6]. The early recognition of this effect was expressed as the "iN-Out" rule of thumb, which explains that in a plasma environment, oxygen has a high tendency to be removed from a molecule, but nitrogen has a tendency to remain in the

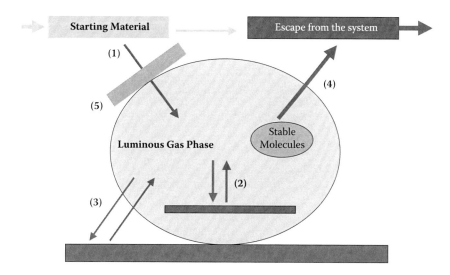

FIGURE 7.4 (*See color insert.*)
The CAP principle: (1) dissociation (ablation) of the monomer to form reactive species; (2) deposition of plasma polymer and ablation of solid, including plasma polymer deposition; (3) deposition to and ablation from nonsubstrate surfaces; (4) removal of stable molecules from the system; and (5) the zone in which dissociation of monomer occurs.

molecule. This rule was originally outlined in plasma treatment of polymers; however, a similar rule seems to apply to the fragmentation of gas molecules in the luminous gas phase.

The electron-impact reactions of an organic molecule do not follow the same process applicable to a simple mono-atomic gas such as Ar. The bond energies that link elements are much smaller than the ionization energies of mono-atomic gases used as carrier gases in low-temperature plasma processes (not plasma polymerization). Consequently, the first step to creating a luminous gas phase, the dielectric breakdown of organic vapor, is the dissociation process as described in Chapter 5. In order to elucidate the mechanisms by which a polymeric material deposits and also those by which surface modification of a polymeric material by plasma proceeds, comprehension of the CAP principle and the plasma sensitivity series seems to be vitally important. Tables 7.4 and 7.5 list bond dissociation energies.

The schematic diagram of the CAP principle is shown in Figure 7.4. In this scheme, there are five major processes necessary to complete the mass balance in the reactor:

1. Dissociation (ablation) of monomer to form reactive species
2. Deposition of plasma polymer and ablation of solid including plasma polymer deposition
3. Deposition to and ablation from nonsubstrate surfaces

TABLE 7.4

List of Bond Energies

Standard Bond Energies

Single Bonds	$\Delta H^{\circ a}$	Single Bonds	$\Delta H^{\circ a}$	Multiple Bonds	$\Delta H^{\circ a}$
H–H	104.2	B–F	150	C=C	146
C–C	83	B–O	125	N=N	109
N–N	38.4	C–N	73	O=O	119
O–O	35	N–CO	86	C=N	147
F–F	36.6	C–O	85.5	C=O (CO$_2$)	192
Si–Si	52	O–CO	110	C=O (aldehyde)	177
P–P	50	C–S	65	C=O (ketone)	178
S–S	54	C–F	116	C=O (ester)	179
Cl–Cl	58	C–Cl	81	C=O (amide)	179
Br–Br	46	C–Br	68	C=O (halide)	177
I–I	36	C–I	51	C=S (CS$_2$)	138
H–C	99	C–B	90	N=O (HONO)	143
H–N	93	C–Si	76	P=O (POCl$_3$)	110
H–O	111	C–P	70	P=S (PSCl$_3$)	70
H–F	135	N–O	55	S=O (SO$_2$)	128
H–Cl	103	S–O	87	S=O (DMSO)	93
H–Br	87.5	Si–F	135	P=P	84
H–I	71	Si–Cl	90	P≡P	117
H–B	90	Si–O	110	C≡O	258
H–S	81	P–Cl	79	C≡C	200
H–Si	75	P–Br	65	N≡N	226
H–P	77	P–O	90	C≡N	213

Source: Organic Chemistry, Michigan State University

[a] Average bond dissociation enthalpies in kcal per mole. (There can be considerable variability in some of these values.)

4. Removal of stable molecules from the system
5. The energy transfer in the zone where dissociation of monomer occurs

Material deposition occurs via plasma formation of reactive species; however, it is not a simple step of forming a polymeric material from a set of reactive species as described in Chapter 5. The reactive species do not necessarily originate from the monomer, because the ablation process could yield reactive species. Gaseous reactive species can originate from once-deposited material (plasma polymer) and also from the reactor wall or any other solid surfaces that are in contact with the luminous gas. A typical example of this case could be found in the presence of N (detected by x-ray

TABLE 7.5

List of Bond Dissociation Energies

Bond Dissociation Energies[a]

Atom or Group	Methyl	Ethyl	i-Propyl	t-Butyl	Phenyl	Benzyl	Allyl	Acetyl	Vinyl
H	103	98	95	93	110	85	88	87	112
F	110	110	109		124	94		119	
Cl	85	82	81	80	95	68	70	82	90
Br	71	70	69	66	79	55	56	68	80
I	57	54	54	51	64	40	42	51	
OH	93	94	92	91	111	79	82	107	
NH_2	87	87	86	85	104	72	75	95	
CN	116	114	112		128	100			128
CH_3	88	85	84	81	101	73	75	81	98
C_2H_5	85	82	81	78	99	71	72	78	95
$(CH_3)_2CH$	84	81	79	74	97	70	71	76	93
$(CH_3)_3C$	81	78	74	68	94	67	67		89
C_6H_5	101	99	97	94	110	83	87	93	108
$C_6H_5CH_2$	73	71	70	67	83	59	59	63	81

[a] In kcal per mole.

Source: Organic Chemistry, Michigan State University

photoelectron spectroscopy, XPS) on the surface plasma deposited material, while no N_2 gas was added to the plasma polymerization system, and the substrate used contains no N-containing moiety. N-atom incorporation onto plasma polymer is caused by reaction (3) in Figure 7.4, deposition to and ablation from nonsubstrate surfaces. In other words, N atom was incorporated into the plasma polymers due to the contamination of the reactor in previous operations, in which N-containing monomer, such as amines and nitriles, were used [7]. The possibility of N atom from N-containing polymeric substrate is very small according to the iN/Out rule.

Important factors described in Figure 7.4 are as follows:

1. *Ablation*, which is fragmentation or dissociation, is involved in every process (i.e., monomer to reactive species, plasma polymer to reactive species, wall surface to reactive species, and escape of fragmented species from the system). Fragmentation of molecules is the primary effect of plasma exposure to a material. The importance of ablation can be visualized in the well-established fragmentation patterns of many organic materials, which constitute the foundation of secondary ion mass spectroscopy (SIMS).

2. The reactive intermediate species are created not only by fragmentation of the monomer but also by fragmentation of the plasma

polymer formed and of materials existing on the various surfaces that come into contact with the plasma.

3. The gases being pumped out consist of "non-polymer-forming" stable species and some unreacted monomers depending on system conditions, such as power input level, flow rate, flow pattern, pumping rate, and shape and size of reactor.

4. The species that do not contribute to polymer formation are basically stable molecules, such as H_2, HF, and SiF_4. When these species are created in the luminous gas phase, the balance between deposition and ablation could shift significantly. However, the manner in which the balance shifts is dependent on the specific system. The formation of stable gases to be pumped out is crucially important in determining the nature of the depositing materials. Some examples, which illustrate the factors described above, are given in the following sections.

7.4 Influence of Unaccounted Factors

7.4.1 Substrate Material

The influence of the CAP effect due to factors other than operation of plasma polymerization could be best illustrated by the following example. An attempt was made to create a fabric with one hydrophilic side and one hydrophobic side by applying a different kind of plasma surface modification to each side [8]. Air plasma treatment was used to make one surface hydrophilic, and CF_4 plasma treatment was used to make the other hydrophobic. Such a fabric with a different set of surface characteristics on each side can be made; however, the success of this undertaking is dependent on which treatment is applied first. The sequence dependency of plasma treatments may be explained by the concept of plasma sensitivity of the elements involved in the two steps. Results are summarized in Tables 7.6 and 7.7.

Examining these results, the following factors involved in this experiment should be kept in mind:

1. Fabrics are porous, and consequently the plasma treatment applied to one side of a fabric penetrates to the other side, even though the second side is not exposed to the plasma directly. This penetration effect of plasma treatment was previously known [12].

2. Likewise, a second plasma treatment will influence the effect of the first plasma treatment, even though the surface treated first is not directly exposed to the second plasma treatment.

TABLE 7.6

Consecutive Treatments of Two Sides of a Polyethylene Terephthalate (PET) Fabric by Different Plasmas[a]

		Contact Angle of Water		
Side	First Treatment	Contact Angle of H_2O	Second Treatment	Contact Angle of H_2O
A	Air Plasma	Water penetrates[b]		58.6°[c]
B		Water penetrates	CF_4 plasma	111.1°
A	CF_4 Plasma	112.7°		Water penetrates[b]
B		54.5°	Air plasma	Water penetrates[b]

[a] Plasma treatment penetrates through a porous structure (i.e., it is not limited to the exposed surface).
[b] Contact angle cannot be measured because the water droplet penetrates quickly into the fabric.
[c] Contact angle can be measured, but water droplet penetrates slowly into the fabric.

TABLE 7.7

Consecutive Treatment of Two Sides of a Polyethylene Terephthalate (PET) Fabric by Different Plasmas[a]

		XPS F Counts		
Side	First Treatment	XPS F Counts	Second Treatment	XPS F Counts
A	Air Plasma	4092		74375[c]
B		4542	CF_4 plasma	63555
A	CF_4 plasma	56376		53921[b]
B		68408	Air plasma	24802[b]

[a] Plasma treatment penetrates through a porous structure (i.e., it is not limited to the exposed surface).
[b] Contact angle cannot be measured, because the water droplet penetrates quickly into the fabric.
[c] Contact angle can be measured, but water droplet penetrates slowly into the fabric.

As is evident in Table 7.6, two characteristically different sides can be obtained only when the air plasma treatment is applied first. This can be explained by the plasma sensitivity of the elements involved. Fluorine is the most electronegative element involved in this experiment. Consequently, it can be removed relatively easily by exposure to the plasma, which consists of less-plasma-sensitive elements (elements of lower electronegativity).

When the two steps of plasma treatment are carried out successively in the same reactor without copious cleaning of the reactor between the two treatments, the influence of the wall surface on the second plasma treatment also becomes an important factor. Table 7.7 reveals the following important facts:

1. Fluorine contamination of the reactor wall is clearly evident. Air plasma treatment introduced a small but clearly identifiable amount of F on the treated surface.

2. The efficiency of F incorporation into the surface structure is higher on the surface that is not directly exposed to the CF_4 plasma. This trend is found consistently in the data presented in Table 7.7.

These findings clearly show that the principle of CAP also applies to glow discharge treatment, which does not intend to deposit plasma polymers, in a similar manner with respect to the interaction of luminous gas with materials.

7.4.2 Reactor Wall Contamination

Reactor wall contamination is one of most annoying factors in plasma polymerization. An obvious effect is the deposition of plasma polymer on unwanted places, which reduces the yield of deposition on substrate. This effect largely depends on reactor design, particularly choice of power source and the expansion of luminous gas phase beyond the substrate surfaces on which plasma polymer coating is attempted. A not so obvious but more serious problem is the change of chemistry of polymer formation mechanisms by the plasma interaction with the materials deposited on the reactor wall during previous operations using the same reactor with different gases. A typical example could be found in the problem encountered in interface engineering for improvement of corrosion protection of aluminum alloy for aircraft by multiple-step plasma processes, when one of the sequential steps was inadvertently skipped. This example also shows, on the other hand, what we could achieve through plasma interface engineering if the designed operation process is strictly followed.

In corrosion protection of metals, in general, one of the key issues is how to handle oxides that cover the top surface of a metal to be treated. In the case of corrosion protection of steel, of which an oxide layer could be removed with relative ease, plasma interface engineering worked very well by removing oxides with hydrogen plasma and depositing Si-containing plasma polymer to improve adhesion of the protective coating [13–16]. With aluminum alloys, however, oxides cannot be removed by hydrogen plasma treatment. The prolonged treatment caused the change of alloy component at the top surface due to the increase of surface temperature, which caused catastrophic damage to the corrosion protection. Consequently, wet chemical removal of oxides was used prior to the plasma interface engineering.

It is important to note that plasma deposition occurs predominantly on the cathode surface in the cathodic polymerization, and a new cathode (substrate) was used in every plasma-coating operation. In other words, the contamination of the reactor is considered to be minimal. The sequence of the three steps of plasma interface engineering is (1) O_2 plasma treatment of a chemically deoxidized aluminum alloy plate, (2) plasma deposition of trimethylsilane (TMS), and (3) plasma deposition of hexafluoroethane (HFE) (i.e., –O_2/TMS/HFE–). This plasma interface engineering yielded excellent adhesion of primer coating applied and excellent corrosion protection. One inadvertent omission of O_2 plasma treatment caused catastrophic failure of the interface engineering due to wall contamination, which could not be cleaned easily.

A particular point needs to be made regarding the application cycle of the plasma processes. In the general scheme of the progression of process steps, ignoring venting and substrate replacement, the O_2 plasma treatment followed the HFE plasma treatment of the previous run, and HFE plasma treatment followed TMS plasma application in the regular operation (i.e., –HFE (in previous run)/O_2/TMS/HFE/O_2 (in the next run)–). Once the O_2 plasma treatment was inadvertently omitted, the order of the sequential plasma processes changed (i.e., the plasma polymerization of TMS followed the HFE plasma treatment, which was performed in the preceding run, so –HFE/TMS/HFE–).

This reversal of the sequence from HFE/O_2/TMS/HFE to HFE/TMS/HFE is the most important issue presented here. The second major issue is the effect of the O_2 plasma treatment on the fluorine-containing contaminants. The XPS analysis of an initial sample revealed virtually no silicon on the alloy surface beneath the lifted primer, but it did indicate a rather substantial fluorine presence. The appearance of a strong silicon signal on the interface side of the removed primer indicated that the entire plasma film had likely delaminated at the interface with the alloy. Analysis of additional samples confirmed that the entire film and primer system delaminated from the alloy panels. The detail of this plasma interface engineering can be seen in references [10,11].

The decrease of Si due to F-containing contaminants and the role of the oxygen plasma treatment can be explained by the principle of CAP. The key factor to explain the change of elementary composition at the interface is the plasma sensitivity of elements involved on the surface and in the plasma phase. The ablation of materials exposed to plasmas appears to follow the plasma sensitivity series of the elements involved, which is in the order of the electronegativity of the elements (i.e., elements with higher electronegativity in the condensed phase are more prone to ablate in plasma that contains elements with lower electronegativity) [8].

When the thin layer of F-containing oligomers is exposed to the TMS plasma, some of F-atoms are removed from the layer and form Si-F moieties in the plasma. Fluorine atoms (high electronegativity) in the contaminants are easily detached from the surface by the interaction of plasma of low electronegative Si (TMS), forming stable species (with Si-F bonds) in the plasma phase, which will be pumped out of the system.

Oxygen plasma treatment of the Al alloy surface with F-containing oligomers is a similar situation but has different consequences, because oxygen plasma does not form polymeric deposition. Oxygen (lower electronegativity than F) plasma ablates F-containing oligomers from the substrate surface. In plasma phase, F-atoms are detached from the organic moieties and become F-containing plasma, which reacts with elements with lower electronegativity in the condensed phase, such as O in metal oxides. Thus, plasma-sensitive F-containing oligomers are converted to more stable (in plasma environment) F-containing inorganic compounds, such as aluminum fluoride-oxide, aluminum fluoride, and so forth, although the details of species are not well known. The reduced concentration of plasma-ablateable F at the surface virtually eliminates F-containing oligomers interfering with TMS polymerization, although F increases on the surface after the oxygen plasma treatment.

The following experiments were performed to confirm the major principles being discussed above: examination of the interference of TMS plasma polymerization by fluorine ablating from the substrate surface, and elimination of plasma-ablateable F by O_2 plasma pretreatment of the substrate surface. First, a very thin layer (less than few nm) of (HFE + H_2) plasma polymer was deposited on an aluminum sheet by 5 seconds of plasma polymerization. One sample was prepared by depositing TMS plasma polymer directly onto this surface. The second sample was prepared by treating the (HFE + H_2) plasma-modified surface with oxygen plasma before depositing TMS plasma polymer. Without oxygen plasma treatment, the TMS plasma polymer did not adhere to the substrate surface, whereas with the oxygen plasma treatment, excellent adhesion was obtained. These results confirm the mechanisms by which the wall contamination caused catastrophic damage to the adhesion of film formed by the plasma CVD process. These summaries are schematically shown in Figures 7.5 and 7.6.

Once a reactor is contaminated to a certain level, it is very difficult to get rid of the persistent influence of the contaminants. During an arbitrarily chosen 1-month period, a contaminated reactor was used only for the following two kinds of samples in order to investigate the decay characteristics of F contamination: one with a plasma-treated new sample plate, and another without plasma treatment. Figure 7.7 depicts the persistence of the contamination examined by following the XPS fluorine contents on the Al alloy surfaces as a function of time. The plasma-treated sample showed considerable levels of F contamination that decreased with the pumping time of the reactor, while untreated samples did not show this effect. The following conclusion could be drawn from the results shown:

1. O_2 plasma treatment incorporates more fluorine (approximately five times) on the Al alloy panel than on the sample just loaded to the reactor and pumped down.

2. XPS studies on the panels prepared with the HFE line disconnected, the liquid N_2 trap, and the vacuum pump oil change confirmed that these are not possible sources of fluorine contamination. The

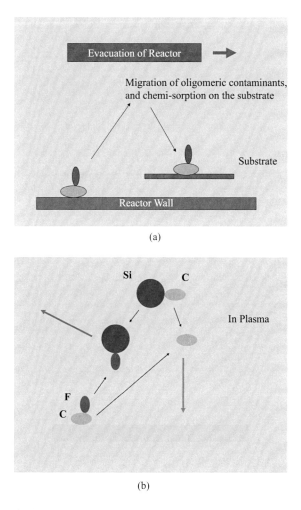

(a)

(b)

FIGURE 7.5 (*See color insert.*)
How F-containing contaminant interferes with plasma polymerization of TMS: (a) migration of
F-containing oligomers; and (b) the interference of TMS deposition by F-containing moieties.

F-content gradually decreases with the evacuation time and intermittent O_2 plasma discharges.

3. From the gradual decrease of the level of contamination with evacuation time with intermittent O_2 or Ar processing, it is considered that
the fluorine level in the regular operation never reached the level
shown at the end of 30 days in Figure 7.7. This is because the HFE/O_2
sequence, which was maintained in the regular operation before the
inadvertent omissions occurred, served as a cleaning process for
the contaminant in the reactor.

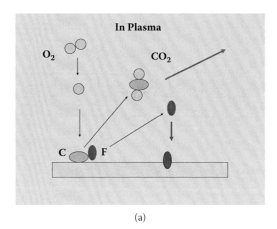

(a)

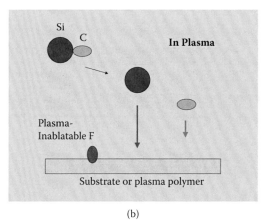

(b)

FIGURE 7.6 (*See color insert.*)
How oxygen plasma treatment prevents the interference of plasma polymerization of TMS by
F-containing oligomers: (a) oxygen plasma treatment; and (b) plasma polymerization of TMS.

It is important to emphasize that O_2 plasma treatment collects much more
F-containing contaminants on the alloy surface than without O_2 plasma treatment, although this treatment virtually eliminated the interference of the
contaminants to the subsequent TMS deposition. In other words, O_2 plasma
treatment does not reduce the amount of fluorine on an aluminum alloy surface, but it reduces plasma-ablateable fluorine on an aluminum alloy surface.

7.5 Dissociation of Monomer Molecules

The extent of dissociation, or fragmentation, of monomer to form reactive species can be visualized by examining the pressure change in a closed-system

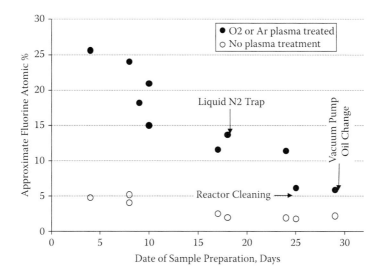

FIGURE 7.7
The decay of fluorine contamination with continued reactor use, involving multiple intermittent chamber evacuations and O_2 or Ar plasma treatments of new substrates, which did not show any effect. Other than this intermittent use and times when chamber-cleaning procedures were employed, the reactor is kept under vacuum, indicating that this contamination does gradually pump out.

plasma polymerization of trimethylsilane (TMS), as shown in Figure 5.9 (Chapter 5). In a closed system, an increase in pressure means an increase in the total number of gaseous species, which can be achieved only by dissociation of the original molecules. In such a system, the composition of the gas phase changes with reaction time. Accordingly, the composition of deposited polymer also changes with reaction time. This change is evident in the XPS profile of polymer deposition from TMS in a closed system, which is shown in Figure 7.8 (the pressure change of which is shown in Figure 5.9).

TMS, $HSi(CH_3)_3$, contains one Si, three C, and ten H in its molecule, and the C/Si ratio is 3. In a flow system, the deposited plasma polymer has a uniform C/Si ratio throughout the film (the lower line in Figure 7.8), indicating the gas phase composition does not change in the steady-state gas flow with reaction time. This also shows that more Si than C is incorporated into the plasma polymer in a flow-system plasma polymerization. In a closed system, the ratio changes with reaction time, reflecting the change in the composition of the gas phase. This also indicates that Si deposits in the early stages of closed-system polymerization, yielding a Si-poor gas phase. The depletion of Si in the luminous gas phase is reflected in the sharp increase of the C/Si ratio in the plasma polymer with increased reaction time. The closed system was run for a longer time than the flow system experiment in order to show the depletion of Si in a closed system.

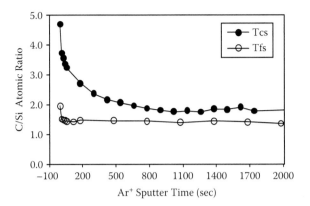

FIGURE 7.8
Depth profile of XPS atomic ratio of C/Si for samples prepared in a closed system (upper line) and in a flow system (lower line).

Data presented in Figure 5.9 (Chapter 5) and Figure 7.8 show how much fragmentation occurs during plasma polymerization and how much influence the fragmentation has on the chemical composition of the resulting plasma polymer. It should be emphasized that the extent of fragmentation is dependent on the various system conditions of plasma polymerization. In the case of deposition onto an electrically floating, conducting substrate and also onto a nonconducting substrate, the energy input level manifested by W/FM is the predominantly important factor. In the case of deposition onto an electrode surface, [(the current density) * p] is the predominantly important parameter, as described in Chapter 5.

The pressure of a flow system plasma polymerization reactor changes when plasma is initiated, if the reactor is not equipped with a throttle valve that is controlled by a pressure gauge. This change depends on the extent of fragmentation of the monomer used, the nature of fragmented molecules, the polymer deposition rate, and the pumping rate of gases. The decrease or increase of system pressure cannot be taken as the tendency or reluctance of the monomer to polymerize. TMS generally causes a large pressure increase; however, its polymerization or deposition rate is much greater than other hydrocarbon monomers that cause large decreases in flow system pressure. The consistently higher deposition rates for Si-containing monomers are partly due to the mass of Si being roughly 2.3 times greater than the mass of C.

7.6 Dependence of Polymer Formation on Operation Parameters

As soon as the luminous gas phase is created, the species that constitute the luminous gas phase, a complex mixture of dissociation products of the monomer molecule, are fed into the reactor. With saturated hydrocarbons and their derivatives, the splitting C-H bond handles the major dissociation reactions. With per-fluorinated compounds, the situation is quite different from the corresponding hydrocarbons, because the bond energy of C-F is larger than that of C-H, which requires the higher energy to break; and the dissociation products (i.e., H_2 and F_2) have opposite trends in their bond dissociation energies. Table 7.4 lists bond energies, and Table 7.5 lists the bond dissociation energies. In order to form polymeric deposition from saturated perfluoro compounds, it is necessary to break C-F bonds to form new C-C bonds, because breaking C-C bonds is counterproductive. The bond dissociation energies of F are consistently greater than the bond dissociation energies as seen in Table 7.5. On the other hand, the bond energy of F-F is only 36.6 kcal/mole, while that of H-H is 104.2 kcal/mole, which means that F is likely to remain as an etching species in the luminous gas phase instead of forming a F-F molecule. Furthermore, under the electron impact dissociation scheme, the electron attachment would form negative ion F^- rather than atomic F, meaning that the probability of forming F-F is very low. Consequently, saturated perfluorocarbons are good etching gases. However, this situation changes when H_2 is added to the plasma-etching system. The bond energy of H-F is 135 kcal/mole, and the addition of H_2 works as a scavenging of etching gas F or F^-. Kay [9] showed that the main effect of CF_4 plasma shifted from etching to polymer deposition when H_2 gas was added, and that the balance between polymerization and etching (ablation) can be controlled by manipulating the amount of H_2.

The double bond in tetrafluoroethylene could avoid the problem described above, because π bond of the double bond is low, 63 kcal/mole, (C=C bond energy) – (C-C bond energy), and the opening of the π bond becomes the major dissociation process. However, in order to plasma polymerize tetrafluoroethylene, it is necessary to keep the energy input level very low, otherwise the competition between plasma polymerization and plasma ablation of the depositing polymer controls the net effect. A good illustration of this situation was presented by Yasuda and Hsu, which is shown in Figure 7.9. The important trend seen in the figure is the decline of the deposition rate with increasing discharge power. At F = 0.5, the deposition rate simply decreases with power. At F = 1.8, the deposition rate increases sharply but reaches the peak and declines sharply more or less linearly with increasing power. At F = 2.6, the higher deposition rates are obtained at low discharge power but a plateau value is reached and then drops very sharply. At F = 4.7, the deposition rate increases sharply at low wattage range and reaches a plateau value. The decline from the plateau value was not observed within the range of discharge power examined, but it is likely to decline at higher wattages.

Figure 7.9 also shows that plasma polymerization cannot be expressed as a simple experimental parameter, such as deposition rate, flow rate, and discharge power. The deposition rate is bound to the flow rate (i.e., the higher deposition rate requires the higher flow rate). The energy input necessary

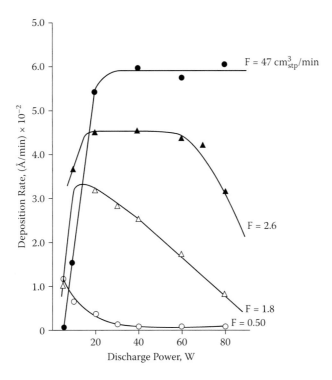

FIGURE 7.9
Dependence of plasma polymer deposition rate of tetrafluoroethylene on discharge power *W* at different flow rates.

to plasma polymerize a monomer gas depends on the mass flow rate. Thus, the deposition rate should be expressed by (deposition rate)/FM, where F is volume or molar flow rate, and M is the molecular weight of gas (i.e., FM is mass flow rate, and discharge power should be given by W/FM). When those parameters are used, the general trend found with hydrocarbon derivatives could be depicted as shown in Figure 7.10. The domains of plasma polymerization can be defined as energy-deficient domain and monomer-deficient domain. Accordingly, the types of plasma polymers could be defined as Type A and Type B plasma polymer. Comparison of plasma polymerization characteristics of F-containing compounds with those for hydrocarbons derivatives without F can be depicted as shown in Figure 7.11. Due to the narrow energy input range permissible, plasma polymerization of F-containing molecules more or less requires multiple bonds and cyclic structure so that di-radical cycle polymerization (Figure 7.1) can be operative. On the other hand, it is interesting to note that (HF_2C-CF_2H) polymerizes fast at low energy input due to the formation of di-radicals by elimination of stable HF from each carbon.

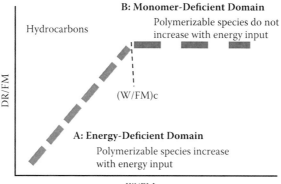

FIGURE 7.10
Deposition characteristics of hydrocarbons in plasma polymerization, and the domains of plasma polymerization.

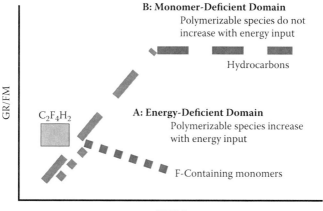

FIGURE 7.11
Comparison of plasma polymerization characteristics of hydrocarbons and F-containing gases.

References

1. Yasuda, H. and C. E. Lamaze, *Journal of Applied Polymer Science*, 17, 1533, 1973.
2. Yasuda, H., *Plasma Polymerization*, Academic Press, San Diego, CA, 1985.
3. Yasuda, H. and T. Hsu, *Journal of Polymer Science: Polymer Chemistry Edition*, 15, 81, 1977.

4. Yasuda, H., *Luminous Chemical Vapor Deposition and Interface Engineering*, CRC Press, Boca Raton, FL, 2004.
5. Yasuda, H., M. O. Bumgarner, and J. J. Hillman, *Journal of Applied Polymer Science*, 19, 531, 1975.
6. Yasuda, T., M. Gazicki, and H. Yasuda, *Journal of Applied Polymer Science. Applied Polymer Symposium*, 38, 201, 1984.
7. Engelman, R. A. and H. K. Yasuda, *Journal of Applied Polymer Science. Applied Polymer Symposium*, 46, 439, 1990.
8. Yasuda, H. and T. Yasuda, *Journal of Polymer Science: Polymer Chemistry Edition*, 38, 943, 2000.
9. Kay, E., invited paper, International Round Table Plasma Polymer Treatment, IUPAC Symposium, Plasma Chemistry, 1977.
10. Moffitt, C. E., C. M. Reddy, Q. S. Yu, D. M. Wieliczka, and H. K. Yasuda, *Applied Surface Science*, 161, (3–4), 481, 2000.
11. Yasuda, H. K., Q. S. Yu, C. M. Reddy, C. E. Moffitt, and D. M. Wieliczka, *Journal of Vacuum Science and Technology. A, Vacuum, Surfaces, and Films*, 19(5), 2074–2082, 2001.
12. Karulkar, P. C. and N. C. Tran, *Journal of Vacuum Science and Technology B*, 3, 889, 1985.
13. Wang, T. F., H. Yasuda, T. J. Lin, and J. A. Antonelli, *Progress in Organic Coatings*, 28, 291, 1996.
14. Chun, H. J., D. L. Cho, T. J. Lin, H. Yasuda, D. J. Yang, and J. A. Antonelli, *Corrosion*, 52(8), 584, 1996.
15. Yasuda, H., T. F. Wang, D. L. Cho, T. J. Lin, and J. A. Antonelli, *Progress in Organic Coatings*, 30, 31, 1997.
16. Lin, T. J., J. Antonelli, D. J. Yang, H. K. Yasuda, and F. T. Wang, *Progress in Organic Coatings*, 31, 351–361, 1997.

8

Operation Parameters
and Deposition Kinetics

8.1 Operation Parameters and Plasma Polymerization Process

8.1.1 Operational Parameter That Influences Repeating
Step Growth Polymerization (RSGP) Mechanism

It is generally considered, quite erroneously, that plasma polymerization starts when an electrical power switch is turned on, and ends when the switch is turned off. In reality, however, actual operation of plasma polymerization starts when the pumping of a reactor is initiated and contains multiple important steps that greatly influence the outcome of plasma polymerization. The steps involved in the plasma polymerization coating depicted in Figure 8.1 [1], as changes of system pressure in each step as a function of operation time, might show the importance of the issue described above. Attention should be paid to the following points with respect to the influence of operational parameters:

1. The overall operation time for the case shown in Figure 8.1 is 50 minutes, in which plasma polymerization time is only 5 minutes.

2. The longest operation time for incremental steps is the evacuation time, 30 minutes, which is more or less arbitrarily determined based on the system pressure. The evacuation time largely depends on what kind of substrates and how much surface area is placed in a plasma polymerization reactor. The influence of substrate to plasma polymerization is a very important factor, for which often less or no attention is paid. It is important to recognize that there is no need for plasma polymerization if we do not need to coat substrates. What kind of plasma polymer should be coated (objective of plasma coating) also entirely depends on the type of substrate. The nature of the substrate and the prehandling before placing it as the substrate for plasma polymerization dictate the most time-consuming step that determines properties of the coating and the value of final products.

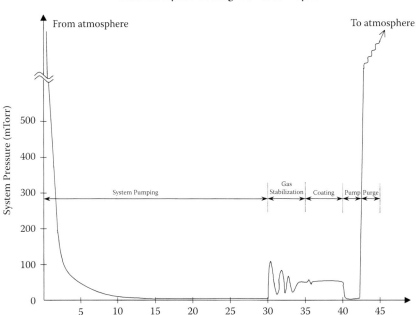

FIGURE 8.1
Overall plasma polymer coating in vacuum. The system pressure changes with process time.

3. If the degassing rate of gases and vapors from the substrate is slower than the pumping rate, the system pressure reaches the minimum pressure that could be read by the pressure gauge employed; however, the low system pressure does not necessarily mean that degassing has been completed. Figure 8.2 [1] depicts the degassing effect of plasma initiation with substrate, which is dried water swollen silicone/hydrogel. Data shown indicate that considerable levels of gases and vapors that are not introduced as processing gas (the mixture of CH_4 and air in the case shown in the figure) were detected. Those gases emanating from the substrate influence the plasma polymerization process and the properties of plasma polymer coating.

4. The level of degassing of substrate and the reactor is a very important factor that influences every important factor of plasma polymerization. When a hydrophilic substrate is used, low system pressure does not mean that degassing is completed, because elimination of H_2O from hydrophilic polymer substrate takes much longer than degassing of dissolved gases. Unfortunately, inception of luminous

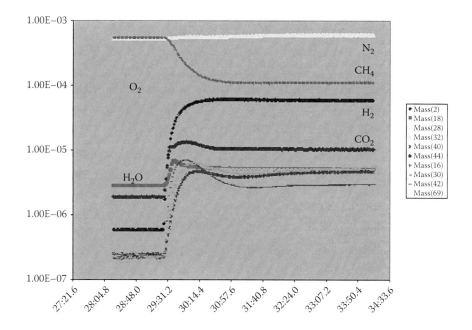

FIGURE 8.2
Change in gas phase components on inception of luminous gas phase.

gas phase releases strongly adsorbed chemical moieties to the luminous gas phase as seen in Figure 8.2.

5. When a substrate that does not contain a hydrophilic component is used in the same operation, the gas phase change due to the inception of the luminous gas phase is much simpler, as shown in Figure 8.3 [1]. The gas phase change is nearly instantaneous, and the hydrogen abstraction as the main chemical reaction to create a chemically reactive species that causes the formation of the depositing species is clearly shown.

Upon completion of evacuation of the polymerization system, a monomer gas is introduced into the system at a predetermined flow rate; however, a steady-state flow of monomer gas does not establish instantly, and some extent of adsorption of gas on reactor surfaces and also absorption of monomer gas by the substrate occurs depending on the nature of gas and substrate material. The gas flow stabilization time is 7 minutes (in the case shown in Figure 8.1), which is longer than the plasma polymerization time of 5 minutes. Consequently, the time the system is exposed to the monomer gas flow is also a very important factor that influences the plasma polymerization as described below.

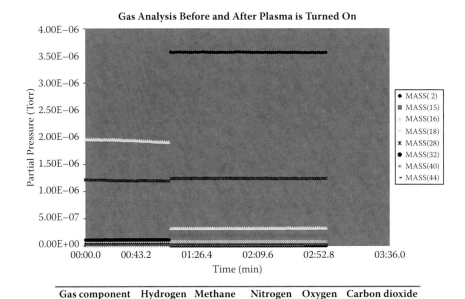

Gas component	Hydrogen	Methane	Nitrogen	Oxygen	Carbon dioxide
Before plasma	0.8%	68.5%	26.3%	2.4%	0.2%
In plasma	65.7%	7.3%	22.9%	0.1%	0.3%

FIGURE 8.3
Gas phase change due to the inception of the luminous gas phase with hydrophobic substrate.

All gaseous species adsorbed or absorbed by materials present in the reactor would come out to the gas phase when the glow discharge is initiated and act as components of the luminous gas phase created. Some gases increase the total amount of gas, due to the dissociation of gas introduced (CH_4, O_2, or N_2) (e.g., dissociation products are H_2, CO_2). A decrease of pressure is due to either deposition (e.g., CH_4) or to being pumped out as reaction products (e.g., O_2).

When the power switch is turned off, the glow discharge ceases, but this does not necessarily mean that plasma polymerization ends at that point, because very reactive free radicals are in the deposited plasma polymer, particularly near the top surface. The monomer gas flow is turned off after the glow discharge was terminated, but generally not at the same time. In this case, the flow of monomer gas sweeps the surface of the plasma polymer deposition for a considerable period of time, which is generally not controlled. Even if the gas flow was stopped at the same time when the electric power was switched off and evacuation of the reactor was started, the evacuation of the monomer gas from the reactor cannot be completed instantaneously but takes some time, which is not controlled or accounted for, and depends on the size and design factors of the reactor and vacuum system. In the case shown in Figure 8.1, this time is about 4 minutes, which is more than 50% of the plasma polymerization time.

With CH_4 as the main deposition gas, which is not a monomer of conventional free-radical polymerization, the effect of this step is negligible. During this unaccounted-for period of time, the free radicals on the surface could react with the monomer gas. If the monomer gas was one of the highly reactive monomers of conventional free-radical polymerization (e.g., acetylene or acrylic acid), graft polymerization by the conventional free-radical polymerization would occur. In such a case, the surface analysis would pick up chemical moieties developed in this unaccounted-for postplasma graft polymerization as the main surface constituent of plasma polymer, which would lead to erroneous interpretation of plasma polymerization, if an effort is made to construct the plasma polymerization mechanism based on the results of such surface analysis.

Real plasma polymerization, not a hypothetical image of plasma polymerization, ends when substrate is taken out of the reactor or exposed to the ambient air in the reactor. Oxygen is the best free-radical scavenger in the ambient gas phase. Thus, oxygen atoms will be incorporated to the surface of plasma polymer deposited on a substrate. This is the main reason why an O atom is found by x-ray photoelectron spectroscopy (XPS) on the plasma polymer of monomers that have no O atom in the molecule. XPS analysis generally shows 15% to 20% O atoms on the surface of plasma polymer of plain hydrocarbons, which do not contain O in the molecules. O_2 mixed with the main deposition gas in magneto-luminous chemical vapor deposition (MLCVD) of CH_4 does not participate with plasma polymerization due to high electron negativity, and O_2 is added for a completely different reason, as discussed in Chapter 6.

Considering all experimental factors that could influence the outcome of plasma polymerization, it is unjustified to consider that plasma polymerization of a specific monomer by a specific method would yield a well-definable plasma polymer of the monomer. In this sense, plasma polymerization of a specific monomer (e.g., acrylic acid or allyl amine) in a specific type of discharge mode (e.g., direct current, DC, or radio frequency, RF, discharge) cannot be reproduced in an unequivocal manner by different reactors in different laboratories, unless the identical reactor and operational procedure are used. Because the system dependency of plasma polymerization is so great, as it can be easily grasped by the RSGP mechanism described in Chapter 7, there is no unequivocally identifiable plasma polymerization of a monomer. This aspect is particularly important with organic molecules that are monomers of conventional free-radical polymerization. On the other hand, this aspect of plasma polymerization can be utilized in an advantageous way to some extent, such as in the case of pulsed-RF discharge to retain some functional groups in a monomer on the deposited plasma polymer. However, it should be kept in mind that material science is based on the principle of compromising material properties, and a gain in one aspect can be obtained at the expense of other features.

8.1.2 Flow Rate and System Pressure of Gas in General

The flow rate of monomer and the system pressure are key parameters of plasma polymerization. The flow rate directly influences the rate of plasma polymerization, because without sufficient monomer in a gas phase in the reactor, plasma polymerization is hindered by the inadequate flow rate of monomer gas. The system pressure has a significant influence on what kind of plasma polymers would result, because system pressure is an important parameter that controls the frequency of gas–gas corrosion, which controls the gas phase reactions that increase the size of reactive species in the luminous gas phase described by the RSGP reactions mechanism. System pressure is also the key parameter to control the ratio of ϕ = (gas–gas collisions)/(gas–surface collisions), which determines the size of depositing species—grain size, so to speak—and adhesion characteristics of depositing species to the substrate surface.

At a given reactor volume and shape of reactor, in which the cross-sectional area of the reactor is large enough with respect to the mean-free pass of gaseous species, ϕ is proportional to system pressure p. In a simplified correlation, the higher the system pressure, the faster is the deposition rate, but the greater is the grain size and the poorer is the adhesion to the substrate. The advantages of MLCVD include this aspect in reversed order (i.e., the lower the system pressure, the slower the deposition rate, but the smaller is the grain size and the better is the adhesion to the substrate, which is discussed in more detail in Chapter 9).

Two parameters, flow rate and system pressure, however, cannot be determined independently in the operation process of plasma polymerization coating. As described in Chapter 7, in glow discharge of Ar, these parameters are not significant factors, because glow discharge in a closed system and that in a flow system do not make any difference, because no gas is consumed by glow discharge. When chemical reactions of gas are involved, particularly plasma polymerization, these factors become crucially important in carrying out plasma polymerization for a specific objective. However, these two factors are not well recognized or their significance is ignored, according to the literature.

Because system pressure of gas is controlled by the product of pressure (p) and the volume of the system (V), the volume flow rate, or molar flow rate (F) of a gas is given by $F = d(pV)/dt$. In most practical cases, the reactor volume is a fixed value (i.e., not a variable parameter of operation); hence, the system pressure of operation is determined by the difference of gas feed-in rate and the gas pump-out rate (i.e., the steady-state system pressure, $p_s = k \{(dp_1/dt) - (dp_2/dt)\}$, where p_1 is the pressure at the high pressure end of gas flow, and p_2 is at the low pressure end of gas flow). The gas feed-in rate, (dp_1/dt), can be controlled by manually opening the needle valve placed on the gas feed-in line, or can be better controlled by an automatic gas flow rate controller, and the gas pump-out rate (dp_2/dt) can be controlled likewise by controlling the pumping-out rate.

8.1.3 Control of Monomer Flow Rate and the System Pressure of Plasma Polymerization Reactor

The control of system pressure of the plasma polymerization reactor requires an additional responsibility of controlling the change, usually an increase of pressure due to the dissociation of monomer gas or a decrease due to the deposition of plasma polymer. With type III monomers, molecules without double bonds, triple bonds, or cyclic structure (see Figure 7.3, Chapter 7), the system pressure increases at the onset of the luminous gas phase. With group I monomers, particularly with triple bond (e.g., acetylene), the deposition rate often is faster than the monomer feed-in rate. In acetylene plasma polymerization at low W/FM level, the deposition of plasma polymer acts as an effective pump, and the system pressure decreases below the transition point pressure, p_θ (see Figure 5.29, Chapter 5), and the glow discharge extinguishes, but when the system pressure increases above p_θ, the glow discharge starts again to lead to a self-pulsating glow discharge polymerization. The control of pumping rate to maintain the system pressure at a constant level eliminates this problem, and, above all, enables us to operate a plasma polymerization under controlled system pressure.

The best way to control the system pressure of the plasma polymerization reactor is as follows:

1. Place a flow controller on the monomer feed-in line, and set the feed-in rate by the delivering rate of gas so that the flow controller will deliver a fixed flow rate of a monomer regardless of the change in system pressure. The flow controller is often called the *mass flow controller*, because the mass-dependent parameter, such as thermal conductivity, is used to control the opening of the metering valve in order to allow a fixed amount of gas that gives the same thermal conductivity of gas in the line. However, the readout of a mass flow controller is converted to volume flow rate (e.g., cc STP/min.), using a conversion factor, often argon, often for only one gas. In order to get volume flow rate of CH_4, it is necessary to find the conversion factor for CH_4, for the controller, because the pressure of gas is dependent on the amount of gas, not the mass of gas according to the gas law. The best way to calibrate the flow rate is to measure the pressure increase in the system, of which the pump-out valve is closed—that is, (dp_2/dt) = 0—and a constant amount

of gas per unit time is fed into the reactor by using the pressure gauge that does not depend on the type of gas (e.g., diaphragm pressure gauge). From the linear slope observed, within a narrow range of pressure increase against time scale, the volume flow rate of the gas can be calculated. The conversion factor for each gas should be similarly determined.

2. Install a diaphragm pressure gauge in the reactor to read the system pressure at the midpoint of the gas flow.

3. Install a throttle valve that can be controlled by the pressure reading of the diaphragm gauge described above. The system pressure during plasma polymerization is controlled by adjusting the throttle valve opening.

With the above settings and operation, a constant flow is established, and the change of pump-out rate is used to control the initial system pressure and maintain the system pressure during plasma polymerization. By this method, it is possible to cope with the change of gas pressure caused by the plasma polymerization process described above. Obviously, the range of system pressure is limited at a given gas feed-in rate, and it might be necessary to pick the appropriate gas feed-in rate to cover the pressure range desired. Without controls of the monomer feed-in rate, the pumping-out rate, and the system pressure of plasma polymerization reactor as described above, plasma polymerization could not be performed in an identifiable and reproducible manner. MLCVD, which is the main subject of this book, is aimed at operating the MLCVD of CH_4 (group 3 monomer) at as low a pressure as possible, and the setup described above works very satisfactorily.

8.2 Deposition Kinetics

8.2.1 Mass Balance in Flow Deposition System

The following parameters are considered *per unit time*:

Total mass of monomer introduced into the system: $W_1 = FM$

Total mass of deposition: W_2

Total mass exits from the reactor: W_3

Monomer-polymer conversion ratio: $Y_p = W_2/W_1$

Total surface area: S

$$W_2 = \oint_S k_1 ds$$

$$S = \oint ds$$

$$k_0 = \frac{k_1}{W_1} = \frac{k_1}{FM}$$

$$\bar{k}_0 = \frac{\oint_S k_0 ds}{\oint ds}$$

$$= \frac{\bar{k}_1}{FM}$$

$$\bar{k}_1 = \frac{\oint_S k_1 ds}{\oint ds}$$

where k_1 is the local mass deposition rate (kg/m^2 s); $\bar{k}_1$ is the average deposition rate; k_0 is the normalized deposition rate, (1/m^2); and $\bar{k}_0$ is the average normalized deposition rate.

The unit of the normalized deposition rate is m^{-2}, and the normalized deposition rate decreases with the total surface, on which deposition occurs. This aspect can be conceived as the "loading factor" of luminous chemical vapor deposition (LCVD). Mass balance in a reactor (flow system) can be established as

$$W_1 = W_2 + W_3.$$

The material formation in LCVD generally requires the production of gaseous byproducts, which do not form deposition, in order to create new chemical bonds for the material formation. For instance, LCVD of saturated hydrocarbons requires hydrogen abstraction in the dissociation glow. In the presence of double and triple bonds, the hydrogen production becomes very small in the low-pressure domain.

Polymer yield, given by $Y_p = W_2/W_1$, cannot be unity because of the gas formation for which gas yield can be defined by $Y_g = W_3/W_1$. The values of Y_p and Y_g can be determined by measurement of pressure of a closed system, which is subjected to electrical discharge. If thus obtained Y_p can be considered as a physicochemical parameter that is characteristic to a specific monomer, the average specific deposition rate can be expressed by

$$\bar{k}_0 = Y_p W_1 \left(\frac{1}{S}\right),$$

where the first term, Y_p, is the monomer characteristic; the second term, mass flow rate, is an operational parameter; and the third term $(1/S)$ is a factor that depends on the reactor design and the loading factor of operation. In other words, the average normalized deposition rate or the average deposition rate obtained by plasma polymerization is a function of the monomer, operational parameter, and design factor of the reactor employed. That is,

Deposition rate = F (monomer characteristics, operational parameter, design factor).

Accordingly, the deposition rate, or deposition characteristic in general, cannot be dealt as a sole function of the monomer or operational parameters such as flow rate and discharge power. This aspect is recognized as the *system dependent* nature of LCVD, which is a very important factor that should not be ignored. Without recognizing the system-dependent nature of plasma polymerization, attempts to scale up a laboratory-scale operation to larger scales often encounter insurmountable difficulties, which leads to failures of large-scale industrial plasma polymerization coatings.

Because the dissociation glow can be considered to be the major medium in which polymerizable species are created, the location of the dissociation glow (i.e., whether on the electrode surface or in the gas phase) has the most significant influence on where most LCVD occurs. It should be reiterated that the dissociation glow is on the cathode surface in DC and audio frequency (AF) discharge but is in the gas phase in RF discharge as shown in Figure 5.5 (Chapter 5). Thus, the deposition pattern is different, and the location of substrate becomes an important factor. The deposition of plasma polymer could be divided into the following two major categories: the deposition that occurs to the substrate placed in the luminous gas phase (*Deposition-G*), and the deposition onto the electrode surface (*Deposition-E*). The partition between *Deposition-G* and *Deposition-E* is an important factor in practical use of LCVD which depends on the mode of operation, which includes the power source.

8.2.2 Normalized Energy Input Parameter to Luminous Gas Phase

In dealing with the topics in this chapter, it is vitally important to recognize that the luminous gas phase, in which formation of species that undergo chemical reactions to yield polymeric material, is not appropriate to be recognized as an ionized gas but mainly as photo-emitting, chemically reactive neutral species that count most in plasma polymerization. The view that the low-pressure plasma, regardless of what type of gas, is an ionized gas stems from physical characterization of Ar discharge, which constitutes the major fundamental studies of glow discharge, and it is necessary to recognize that Ar discharge is not the subject of plasma polymerization, because it does not deposit polymeric materials. Fundamental steps involved in the deposition plasma are significantly different from those in Ar plasma as described in previous chapters.

In LCVD (low-pressure plasma polymerization), activation (formation of the reactive species) and deactivation (deposition of materials) occur in the same gas phase, because the power input is directly applied to monomer gases, and the material formation occurs mainly in the same gas phase. The activation of organic molecules in a flow system operation of LCVD occurs mainly in the dissociation glow. The supply of the monomer into the luminous gas phase is a crucially important factor, because the monomer is consumed in the glow by depositing polymers, and the numbers of polymer-forming species in the glow decrease as plasma polymerization progresses.

The most important factor that influences the properties of plasma polymers obtainable from a monomer is the energy input level of the plasma polymerization process. The energy input level determines the extent of dissociation and the level of scrambling atoms in the monomer molecule. The retention of molecular structure of the original monomer decreases with the increasing level of energy input. The energy level in the luminous gas phase can be manifested by the parameter, W/FM, where W is the discharge wattage, F is molar or volume flow rate, and M is molecular weight of the monomer [2]. The parameter has units of J/kg (i.e., energy per mass, of monomer). The numerical value of W/FM in J/kg can be calculated from the input energy, W, in watts, volume, or molar flow rate; F in sccm; and the molecular weight of gas molecule used as a monomer, M, in grams/mole by

$$[W/FM \text{ in J/kg}] = W/FM \times 1.34 \times 10^9.$$

W in watts is the energy input into the electrical discharge system, whereas W/FM is nominal energy input into the luminous gas phase in which LCVD (plasma polymerization) occurs. This subtle but very important difference could be visualized by the analogy in how a 10 W lightbulb can be used as a heat source in a closed system. A 10 W lightbulb consumes 10 W of electrical energy. In this case, 10 W means the energy input into the electrical circuit of the bulb. The influence of the bulb on the surrounding medium, on the other hand, depends on the conditions of the environment. The bulb placed in a room hardly influences the temperature of the room. If the same bulb is placed in a small box used as an incubator, the bulb could be used as the sole heat source to control the temperature of the incubator. If the same bulb is kept on a hand, it could cause a burn. In these cases, the energy input to the bulb, 10 W, has no meaning by itself with respect to the thermal effect to the system in which the bulb is placed. In order to see the effect of the light-bulb on the surrounding system, it is necessary to divide the energy input to the bulb by the total mass of air surrounding the bulb. If the bulb is placed in a flow system, the total mass of air is given by the mass flow rate of air. In the same analogy, W is the energy input to the electrical circuit of the glow discharge generator (with correction of the reflected wattage), in the case of RF discharge, but the electrical energy input to the luminous gas phase, which is given by the product of the voltage V and the current I (i.e., $W = V * I$) in the case of DC and AF discharges. W/FM, in Joule/kg of monomer, represents the energy input to the luminous gas phase in which plasma polymerization occurs. It has been well established that the plasma polymerization that occurs in the luminous gas phase is primarily controlled by a composite power parameter, W/FM [3].

When the deposition rate is measured at a fixed flow rate and varying power input W, a line showing the dependence on W is obtained. However, when the same experiment is carried out at different flow rates, different lines are obtained. Figure 8.4 depicts the interrelated effects of W and F on the deposition rate. At lower flow rates, the deposition rate reaches a plateau value and becomes wattage independent. The critical value of W, at which the leveling off occurs, increases with increasing flow rate.

When the same experiments were carried out as a function of flow rate at various fixed input power W, the deposition rates could be plotted as a function of flow rate as depicted in Figure 8.5. Similar saturation effects are also seen in this graph as a function of flow rate. At low flow rate, however,

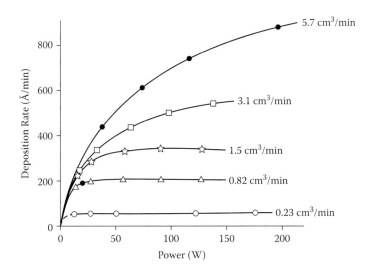

FIGURE 8.4
Dependence of deposition rate of plasma polymer of tetramethyldisiloxane on discharge wattage at various monomer flow rates (cm³/min).

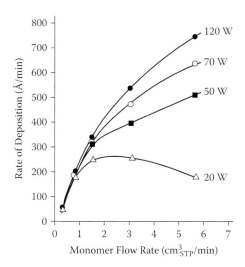

FIGURE 8.5
Dependence of deposition rate of the plasma polymer of tetramethyldisiloxane on the flow rate of the monomer.

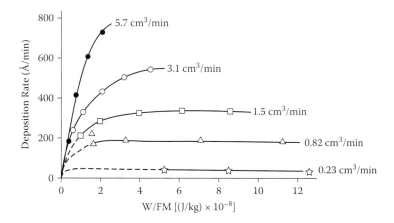

FIGURE 8.6
Dependence of deposition rate of the plasma polymer of tetramethyldisiloxane on W/FM at various monomer flow rates.

a decline in the deposition rate, rather than staying at the plateau value, is observed. The same data shown in Figures 8.4 and 8.5 can be presented in one graph as a function of W/FM, as shown in Figure 8.6.

Based on W/FM, the material formation in LCVD can be divided into two regimes: an energy-deficient regime and a monomer-deficient regime. In the energy-deficient domain, ample monomer is available, but the power input rate is not sufficient. In this domain, the deposition rate increases with the power input. In the monomer-deficient (power-saturated) domain, sufficient discharge power is available, but the monomer feed-in rate is the determining factor for the deposition. In this domain, the deposition rate increases with the flow rate.

The decline of deposition rate with increasing flow rate seen in Figure 8.5 is due to the fact that the increase of flow rate at a fixed W decreases the value of W/FM. In the low W case, the decrease of W/FM crosses the borderline of two domains (i.e., from the energy-deficient domain to the monomer-deficient domain).

It is important to note that these two domains cannot be identified based simply on the value of operational parameters. The domain can be identified only by the dependence of the deposition rate on operational parameters W/FM as depicted in Figure 8.6. The critical value of W/FM at which the domain changes is dependent on the nature of organic molecules as described in Chapter 7.

8.2.3 Normalized Deposition Rate

Most experiments start from the power-deficient domain, where the deposition rate (D.R.) can be expressed by the following expression:

$$D.R. = \alpha W \tag{8.1}$$

Dividing both sides of the equation by FM, one obtains

$$D.R./FM = N.D.R. = \alpha W/FM \tag{8.2}$$

$$N.D.R. = \alpha \ (W/FM) \tag{8.3}$$

where $D.R.$ is the mass deposition rate, and $N.D.R.$ is the normalized deposition rate.

The normalized deposition rate is the only form of deposition rate that can be used to compare deposition characteristics of different monomers with different chemical structures and molecular weights under different discharge conditions (flow rate, system pressure, and discharge power). Similarly, W/FM can be considered as the normalized power input. When only one monomer is employed, $D.R.$ can be used to establish the dependency of deposition rate on operational parameters. Even in such a simple case, $D.R.$ cannot be expressed by a simple function of W or F, and its relationship to those parameters varies depending on the domain of plasma polymerization.

As the power input is increased (at a given flow rate), the domain of plasma polymerization approaches the monomer-deficient one, which can be recognized by the asymptotical approach of $D.R.$ value to a horizontal line as the power input increases. In the monomer-deficient domain, the deposition rate (plateau value) will increase as the flow rate is increased and shows a linear dependence on the monomer feed-in rate at a given discharge power and the system pressure (Figure 8.5); that is,

$$D.R. = \beta \, (FM.) \tag{8.4}$$

This relationship is valid only in the monomer-deficient domain. The further increase of the flow rate (*FM*) will eventually decrease the deposition rate as the domain of plasma polymerization changes to the energy-deficient domain. Equation (8.3) includes data points in such a decreasing part in Figure 8.5. An increase in flow rate (at a given discharge power) has the same effect as decreasing the discharge power (at a given flow rate); conversely, an increase of discharge power has the same effect as decreasing the flow rate.

Figure 8.7 illustrates how well the normalized thickness growth rate, *GR/FM*, in 15 kHz and 13.5 MHz LCVD of methane and *n*-butane, can be expressed as a function of the composite input parameter *W/FM* [4]. It is important to recognize that regardless of the mass of monomer, flow rate, and discharge wattage, a single line fits all data obtained in 15 kHz or 13.5 MHz LCVD of hydrocarbons employed, in which the deposition occurs on an electrically floating conductor or on a dielectric substrate placed in the glow. However, the line for 15 kHz and the line for 13.5 MHz are not the same. Although these general trends are valid and important, the value of *GR/FM* as well as *W/FM* are not unique; an infinite combination of *GR* and *FM*, as

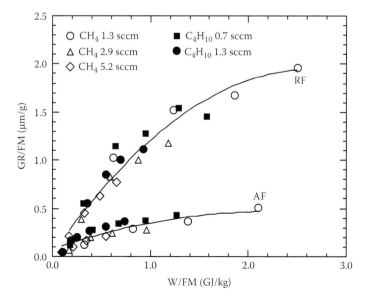

FIGURE 8.7

Dependence of *GR/FM* on *W/FM* for 15 kHz (audio frequency, AF) and 13.56 MHz (radio frequency, RF) magnetron discharge (flow rate: methane 1.3, 2.9, 5.2 sccm; *n*-butane 0.7, 1.3 sccm).

well as *W* and *FM*, yields the same value of these parameters. This feature is the difficulty of the "system dependency" of plasma polymerization.

Thus, material formation in the luminous gas phase (*Deposition-G*), which is given in the form of normalized deposition rate (*D.R./FM*), can be controlled by the composite parameter *W/FM* (normalized energy input parameter), which represents the energy per unit mass of gas, J/kg. Because of the system-dependent nature of LCVD, *W/FM* is not a unique parameter and varies depending on the design factor of the reactor. The value of *W/FM* in a reactor might not be reproduced in a different reactor; however, the dependency remains the same for all *Deposition-G*.

The results in Figure 8.7 indicate that the deposition rate by RF discharge is significantly greater than that by 15 kHz discharge. However, this is not a general trend applicable to all cases. The deposition rate on substrate placed in the gas phase depends on the nature and the location of the dissociation glow, which vary depending on the mode of discharge and nature of monomer, as described in more detail in the next section.

The product of (deposition rate) and the deposition time determine the film thickness. Hence, (*W/FM*)*t* is an important practical parameter to control the thickness of the deposition. In many practical applications, in which the actual thickness of deposition is extremely difficult to measure, the overall functional character of LCVD process can be controlled by this parameter, (*W/FM*)*t*, while maintaining a prefixed value of (*W/FM*).

8.3 Properties of Plasma Polymers and Domains of Plasma Polymerization

8.3.1 Type A and Type B Plasma Polymers

In LCVD, polymerization and deposition are inseparable components of the polymerization-deposition mechanism, but these two processes are not coupled. None of the reactions considered in the RSGP scheme shown in Figure 7.1 (Chapter 7) is a polymerization by itself. While repeating the steps, the species vary in size, depending on how many cycles have been progressed, deposit on the substrate surface. The number for how many times the cycles, which are shown in Figure 7.1, repeat before a species deposit can be expressed by the term *kinetic path length*. As the kinetic path length in gas phase increases, the size of the gaseous species increases, and the saturation vapor pressure of the species decreases, and this forces the species to deposit.

The formation of reactive species from the monomer or from the (non-reactive) products of reaction is essentially a destructive process; that is, it requires the breaking of a bond (e.g., C-H, C-F, or C-C, etc.). Consequently, how far these step reactions have progressed before deposition occurs will influence the chemical and physical nature of the polymeric deposit. This situation can be visualized by the temperature dependence of material deposition for LCVD of perfluoro-2-butyltetrahydrofuran (PFBTHF) at different discharge power input, which was obtained by changing only the temperature of a thickness monitor placed in a steady-state glow discharge of the monomer without changing other parameters of glow discharge [5].

In LCVD, the first step to create reactive species is not a thermal process, and the temperature (of substrate surface) dependence of overall material deposition merely reflects the temperature dependence of the deposition step. If the deposition rate k at a different substrate temperature is plotted against $1/T$, it yields negative slope, indicating that the rate process cannot express the deposition rate (i.e., the deposition process is more or less an adsorption or condensation process).

The plasma polymerization reactor used is schematically shown in Figure 8.8. Parallel electrodes by a 10-kHz power source created glow discharge of the monomer. A thickness monitor sensor is placed at the projected circumference of electrodes intercepting the mid-electrode plane. The circulation of a temperature-controlled liquid controlled the temperature of the crystal surface, on which the plasma polymer deposits. In order to measure the substrate temperature accurately, two thermocouples are placed in the fluid-circulating tubes (inlet and outlet) just outside of the plasma reactor. The substrate temperature is estimated from the average of the thermocouple readings. The temperature dependence of plasma polymer deposition is measured by changing the substrate temperature while maintaining all other

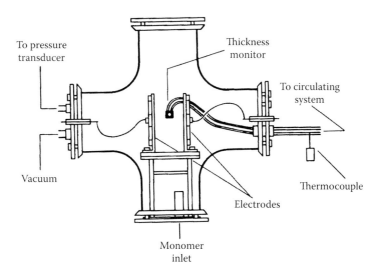

FIGURE 8.8
Reactor used in the temperature dependence study.

operational parameters constant. The plasma polymerization of a monomer is investigated by starting at the highest temperature (80°C).

The significance of these experiments is as follows. In plasma polymerization, the creation of reactive species that cause the formation of plasma polymer, which might be described as the plasma activation step, and the process that causes the deposition of reactive species, which might be termed a plasma deposition step, are not coupled (i.e., these two steps are independent). This is a significant difference from the conventional free radical polymerization, in which the activation step dictates the formation of the polymer, the rate of polymerization and the molecular weight of the polymer. Thus, the activation step and the termination step are coupled in conventional free-radical polymerization. In the experiments described here, the plasma activation step is kept constant, and only the deposition step is altered by changing the surface temperature of the small deposition monitor without influencing the temperature of the overall reactor system.

Figure 8.9 depicts the temperature dependence of the deposition rate for the plasma polymerization of perfluorbutyltetrhydrofuran (PFBTHF)

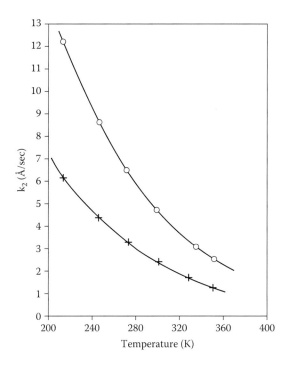

FIGURE 8.9

Temperature dependence of deposition rate k_2 for tetrafluoroethylene. Flow rate (cm$^3_{STP}$ /min) and power (W); O: 1.04 sccm and 9.84W; X: 0.54 sccm and 4.89.

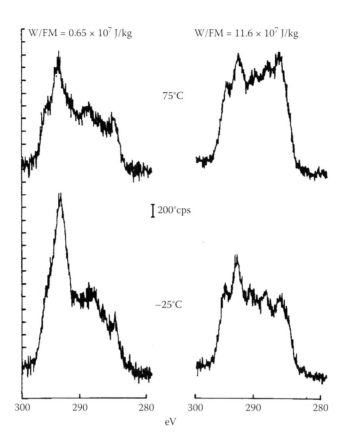

FIGURE 8.10
ESCA C1s spectra of plasma polymers of perfluoro-2-butyltetrahydrofuran obtained at different *W/FM* and substrate temperatures.

shown as plots of the deposition rate constant k versus T. The XPS C_{1S} spectra of polymers deposited at different temperatures under different energy input levels are shown in Figure 8.10. The critically important point, which is not well recognized or understood by researchers in the field of plasma polymerization, is that the formation of reactive species which is tied to the dielectric breakdown of the gas phase and the deposition of reactive species that underwent the step of chemical reaction to increase the size are not coupled. Under such a situation, the species detected in the dissociation glow zone do not have unique influence on the deposition and the characteristics of deposited material. It is an unthinkable situation in conventional free-radical polymerization. However, in the initiated ionic polymerization, both cationic and anionic polymerization, the decoupled initiation and termination are the case, because the termination occurs with impurities [3].

Important aspects of the results are as follows:

1. Temperature dependence is negative, which means that what we observe as the temperature dependence of the polymer deposition is not the reflection of the reaction rate of the growth reactions, which confirms the decoupled plasma activation step and the deposition step.
2. The reaction products obtained at different substrate temperatures are not the same, which means that the kinetic path length (how many cycles has progressed) of depositing species changes the properties of the plasma polymer, which confirms the RSGP mechanism of plasma polymerization.

Data in Figure 8.10 show the effect of energy input and of substrate temperature, but the temperature effect is a clear demonstration of the decoupled activation and the deposition. Once we recognize the overall mechanism of plasma polymerization, we could find the similar effect in what is otherwise considered a routine operation. The following figures show if one could get the same plasma polymer under the same experimental parameter but in different reactors, and also if one could get the same plasma polymer at a different location within a reactor. Figure 8.11 depicts the reactors used and the location of a sample collection for XPS analysis of RF plasma polymerization of perfluoropropene. The ID of a medium-size reactor is 3.3 cm, and of a large-size reactor is 4.7 cm. Figure 8.12 shows C1s XPS spectra of sample collected in the center part of a reactor at the middle of two electrodes. Figure 8.13 shows spectra collected at 15 cm upstream position, and Figure 8.14 at the 15 cm downstream position. These figures show the complexity of plasma polymer formation and the plasma polymer deposition mechanism: no same polymer could be obtained in different reactors or at different positions within the same reactor.

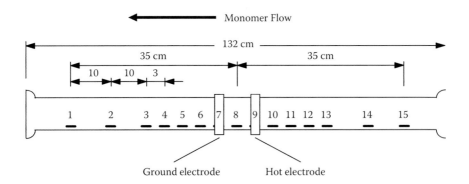

FIGURE 8.11
Location of deposition sample collection in a tubular reactor.

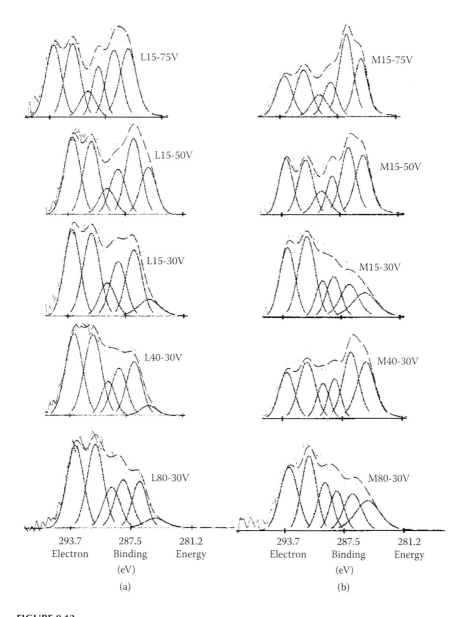

FIGURE 8.12
XPS C1s spectra of perfluoropropene plasma polymer collected at 15 cm upstream position:
(a) large reactor, (b) medium reactor. *W/FM* increases from the top figure to lower figures.

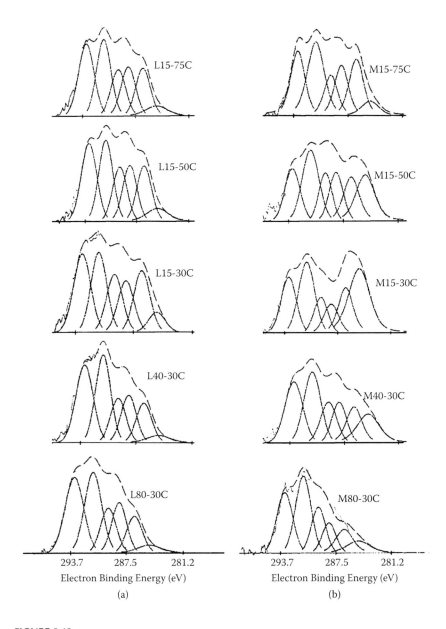

FIGURE 8.13
XPS C1s spectra of perfluoropropene plasma polymer collected at the center position location:
(a) large reactor, (b) medium reactor. *W/FM* increases from the top figure to lower figures.

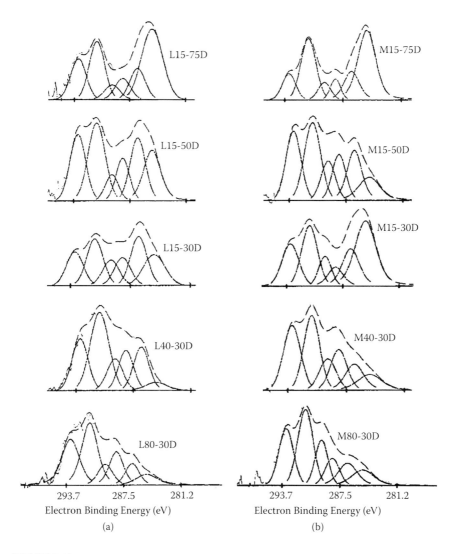

FIGURE 8.14

XPS C1s spectra of perfluoropropene plasma polymer collected at 15 cm downstream position: (a) large reactor, (b) medium reactor. *W/FM* increases from the top figure to lower figures.

We could understand these situations if we refer to the complex scheme of the RSGP mechanisms and also if we recognize the feature of *deposition-growth* polymerization, which is built into the RSGP mechanisms. According to the RSGP mechanisms, the deposition of polymer and the growth reactions cannot be clearly identified, because they are independent steps. In other words, the deposition of material (polymer) occurs neither by the deposition of formed polymers (in gas phase) nor by the discrete process of molecular growth of the deposited monomers. It is important to recognize the following important aspects:

1. Multiple chemically reactive species are created from a monomer in the dissociation process.
2. The kinetic path length changes the chemical structure of the species.
3. As the size of the species increases, as the consequence of reactions in the gas phase, the vapor pressure of the species decreases and enhances the condensation (deposition of reactive species) on the surface.
4. The kinetic path length less than probably five (cycles) would yield large enough molecular size from the smallest reactive species created by the dissociation of a monomer that cannot remain in the gas phase.
5. The deposition of species from the luminous gas phase to the surface does not mean the end of the polymer formation process. It only changes the location of the molecular size increasing process.
6. The ambient temperature of the glow in the typical plasma polymerization reactions, described in this book, is generally in the vicinity of 380 to 400 K and remains reasonably constant after a steady-state condition is established.

Any species involved in the RSGP mechanisms can deposit on a substrate surface. Deposition occurs when an impinging particle fails to bounce back from a colliding surface. Such a deposition may result from the loss of kinetic energy or from the formation of a chemical bond with a target molecule or atom. Thus, the deposition coefficient is a function of the nature of the species involved (mass, kinetic energy, chemical reactivity, etc.) and the surface temperature. What we observe as the temperature of substrate changes is a reflection of the temperature dependence of the deposition coefficients of various species involved in the RSGP mechanisms. The kinetic path length plays a key role in the change of the deposition coefficients.

The activity a of a vapor in a system is given by $a = p/p_0$, where p is the partial pressure of the species, and p_0 is the saturation vapor pressure of the species at the specific temperature of the system. As p_0 decreases as a consequence of increasing size, a approaches unity, where the condensation of the species takes place. In the vapor phase near the substrate (Figure 8.10), which is at a lower temperature, the value of p_0 is lower. Consequently, at a lower substrate temperature, the growing species with shorter kinetic path lengths deposit.

The most important factor, in general cases, that influences the properties of plasma polymers from a monomer is the energy input level of the plasma polymerization process. The energy input level determines the extent of fragmentation or scrambling of the monomer molecule. The molecular nature of plasma polymerization decreases with the increasing level of energy input. The energy level in the diffused plasma can be manifested by the parameter, W/FM, where W is wattage, F is molar or volume flow rate, and M is molecular weight of the monomer. The parameter has units of J/kg (i.e., energy per mass [of monomer]).

8.3.2 Utilities of Type A and Type B Plasma Polymers

Plasma polymerization is system dependent, and a monomer does not yield a well-defined polymer that can be identified by a plasma polymer of the monomer (e.g., plasma polymer tetrafluoroethylene, acrylic acid, etc.). Plasma polymers formed at the high W/FM end of the power-deficient domain and also in the monomer-deficient domain are tight, three-dimensional amorphous networks, which contain less discernible functional groups and are a covalently bonded assembly of atoms C and Si, and so forth, which could form such networks (type B plasma polymers). Plasma polymers formed at the low W/FM end of the power-deficient domain could contain functional groups present in the monomer, but the network structure is much looser and often consists of deposition of oligomers (type A plasma polymer).

The coating that has tight network structure with functional groups cannot be obtained by a single-step plasma polymerization of a monomer, regardless of how operational parameters are manipulated. The retention of functional groups of monomer by some efforts, such as pulsed discharge, remote-plasma deposition, and so forth, can be achieved at the expense of the unique characteristics of the type B plasma polymers, such as good barrier characteristics and good adhesion to the substrate. It seems to be a dream goal of plasma polymerization for some users of the technology to have the coatings that have advantageous characteristics of type B plasma polymer with functional groups in the monomer. It is necessary to choose the type depending on why plasma polymerization should be used.

Plasma polymers formed at the high W/FM end of the power-deficient domain and also in the monomer-deficient domain are tight, three-dimensional amorphous networks that do not contain discernible functional groups (type B plasma polymers). Plasma polymers formed at the low W/FM end of the power-deficient domain could contain functional groups in the monomer, but the structure is much looser and often consists of oligomeric deposition (type A plasma polymer). Type B plasma polymers have unique advantageous features in the practical application of plasma coating (e.g., imparting biocompatible surface, improving corrosion protection, and various interface engineering of materials), but those advantageous features make analytical evaluation of surface and bulk characteristics of a nanofilm much more difficult than those for type A plasma polymers.

Type A plasma polymers have advantages in that analytical evaluations are much more feasible than with the type B plasma polymer. However, the latter advantage of the type A plasma polymer is gained at the expense of practical values obtainable with type B plasma polymers. These trends seem to coincide with the trends that an overwhelming number of publications are with type A plasma polymers. In spite of this trend, the decisive industrial applications of type A plasma polymer seem to be in the category of "lackluster."

8.4 Partition of Deposition on Electrode and Deposition on Surface in Gas Phase

8.4.1 Cathodic Polymerization versus Polymerization in Negative Glow of Direct Current (DC) Discharge

The real significance of DC cathodic polymerization, in spite of relatively limited applications of the method, lies in the fact that the dissociation glow of deposition gas exists as the cathode glows (i.e., the dissociation glow is not in the gas phase, and it is possible to follow the path of plasma polymerization/deposition as a function of the distance from the cathode surface to the substrate surface, which is a unique feature of DC cathodic polymerization). Because of the fact that the dissociation glow in DC discharge exists as the cathode glows, DC discharge is an ideal method to deposit plasma-polymerized nanofilm on a metal surface, which is used as the cathode. On the other hand, the same feature prevents DC discharge in deposition of plasma polymer on substrate placed in the gas phase, because DC discharge seizes when the thickness of deposition reaches a certain level; the cathode becomes insulated by the deposition.

In the widely used RF discharge, the dissociation glow is in the gas phase, and because of that, the dissociation glow in RF discharge has been intuitively assumed to be the ionization glow according to the image carried over from the glow of Ar DC discharge, although the distinction of dissociation glow and ionization glow did not exist before the discovery of the dissociation cathode glow in recent years [6–10]. Before the recognition of dissociation glow as the cathode glow, the deposition characteristics of DC plasma polymerization appeared as if DC plasma polymerization occurs in the cathodic dark space, and the polymerization was described as "dark (space) plasma polymerization" [11]. In this chapter, characteristics of DC cathodic polymerization are examined in comparative manner with alternating current discharges: 40 kHz and 13.5 MHz RF discharge.

8.4.2 Deposition Profile on Electrode

In DC cathodic polymerization, the activation of reactive species and deposition of polymers mainly occur in the cathode glow (molecular dissociation glow, which touches the cathode surface). In a glow discharge initiated by an alternating current power source (e.g., 40 kHz), the electrode can be visualized as an alternating cathode in half of the discharge time and anode in the other half. In RF, the oscillating electrons in the glow discharge are mainly responsible for the creation of polymer-forming species. As a result, in RF discharge, the molecular dissociation glow no longer touches the electrode surface but remains very close to the electrode surface. Consequently, the role of electrode, with respect to the deposition onto the electrode, changes dramatically.

As described in previous chapters, the material formation in LCVD takes place mainly in the molecular dissociation glow. In a diffused luminous gas phase such as a relatively large-volume RF discharge, the contributions of the electron-impact dissociation are difficult to identify. In DC discharge, the chemical reactions in the molecular dissociation glow are dominant, and the majority of deposition occurs to the cathode surface. Accordingly, the analysis of deposition onto the electrode surface would provide important information for comprehending the difference due to the nature of the power source.

Comparing the deposition rate dependence on operational parameters for the deposition in the diffused luminous gas phase and for the cathodic deposition, the contribution of the cathodic polymerization can be estimated by examining the system pressure dependence of the deposition rate (at a fixed flow rate). If the material formation in the diffused luminous gas phase is the dominant factor, it is anticipated that the deposition rate would be independent of the system pressure. If the material formation in the molecular dissociation glow is the dominant factor in a discharge created by an alternating polarity power source (AF discharge), the deposition rate onto an electrode is dependent on the system pressure, and the value of the deposition rate is expected to be one half of that for DC discharge.

Figure 8.15 depicts the two configurations of the electrode–substrate arrangement used for TMS deposition in different glow discharges [11–13]. In configuration (a), the Al substrate was used as the powered electrode, which is the cathode in the DC process. In such a configuration, the two parallel stainless steel plates were used as grounded electrodes, which are on the same electrical potential. In configuration (b), the Al panel was used as a floating substrate positioned in between the two parallel electrodes. In this configuration, one of the two parallel electrode plates was used as the powered electrode (or cathode in DC process), and another was used as the grounded electrode (or the grounded anode in DC process).

When the equation for plasma polymerization (Eq. 8.2) is applied to express the thickness growth rate of the material that deposits on the cathode, *cathodic polymerization*, it becomes quite clear that the deposition kinetics for the cathodic polymerization is quite different. There is a clear dependence of the deposition rate on W/FM, but no universal curve could be obtained. In other words, the relationship given by Eq. (8.2) does not apply to *cathodic polymerization*. The best universal dependency for *cathodic polymerization*

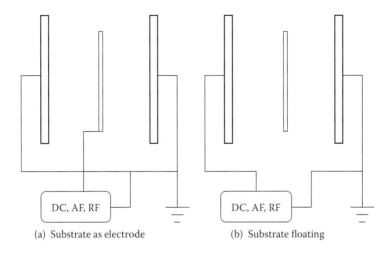

FIGURE 8.15

Two configurations of the electrode setup used in the glow discharge polymerization process: (a) Al substrate was used as powered electrode, and (b) Al substrate was floating in between the two parallel electrodes.

was found between *D.R./M* (cf., not *D.R./FM*) and the current density (*I/S*), where *I* is the discharge current, and *S* is the area of cathode surface [6]. Figure 8.16 depicts this relationship for all *cathodic polymerization* data, which were obtained in the same study, covering experimental parameters such as flow rate, size of cathode, and mass of hydrocarbon monomers but at a fixed system pressure.

The implications of the correlation shown in Figure 8.16 are as follows:

1. The energy input parameter (based on the luminous gas phase) does not control the deposition of material onto the cathode surface.
2. The current density of a DC glow discharge is the primary operational parameter.
3. The flow rate of monomer does not influence the film thickness growth rate.
4. The film thickness growth rate is dependent on the mass concentration of monomer (c_M) in the cathode region rather than the mass input rate (*FM*). (In these experiments, the system pressure was maintained at a constant value of 50 mtorr; thus, *c* was a constant.)

The cathodic deposition, in general cases, can be expressed by the following equation:

$$D.R./[M] = \alpha_c [I/S] [c_M] \tag{8.5}$$

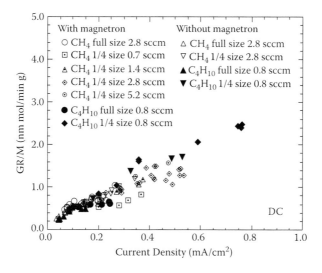

FIGURE 8.16

A master curve for the relationship between GR/M and the current density for DC cathodic polymerization. Data were obtained under various conditions for methane and *n*-butane, at a fixed system pressure of 50 mtorr.

Because the variation of voltage is small in wattage-mode operation, $[I]$ could be replaced by $[W]$ for practical comparison,

$$D.R./[M] = \alpha_c [W/S] [c_M] \qquad (8.6)$$

The concentration of gas c is given by $c = p/RT$, and deposition rate can be given in a similar manner as Eq. (8.3).

$$D.R./[M] = \alpha' [W/S] [p] \qquad (8.7)$$

The equation indicates that cathodic polymerization is controlled by the conditions of the local environment near the cathode. The normalized deposition rate in DC *Deposition-E* is $D.R./[M]$, not $D.R./[FM]$, and the normalized power input parameter is Wc_M/S, not W/FM. In DC discharge, the dissociation glow virtually adheres to the cathode surface. Therefore, the equation proves that the dissociation glow controls the deposition rate on the cathode surface.

The fact that DC cathodic polymerization depends on only the mass concentration near the cathode (dissociation glow) is strong evidence that the monomer gas is drawn into the dissociation glow due to the decrease of partial pressure of the monomer gas in the dissociation glow. This effect is clearly observable in DC cathodic polymerization, because the dissociation glow is the cathodic glow. This effect is less conspicuous in RF discharge,

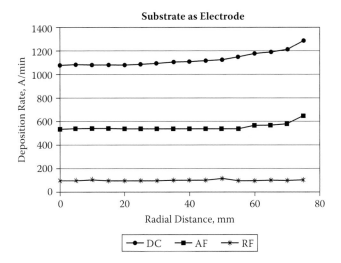

FIGURE 8.17
The deposition rate profiles of TMS in DC, 40 kHz, and 13.56 MHz LCVD with substrate as electrode; 1 sccm TMS, 50 mtorr system pressure, 5 W power input.

because the dissociation glow is spread out wider due to the different mode of electron-impact dissociation; however, it is reasonable to consider that the major dissociation takes place in the strongest glow near the energy input surface, which depends on the mode of RF energy coupling to the gas phase.

Trimethylsilane (TMS) deposition rate profiles in DC, 40 kHz, and 13.56 MHz discharges are shown for electrodes in Figure 8.17. It can be seen that, regardless of the frequency of electrical power source used, a uniform deposition of TMS polymers was observed in the three plasma processes, although an appreciable edge effect occurred in the DC, and a less pronounced effect occurred in the 40 kHz when the substrate was used as the cathode or powered electrode. No edge effect is found with RF discharge, because there is no electric field, as exists in DC and 40 kHz discharge, in RF discharge. The uniform distribution of deposition rates justifies the use of single measurement at the center of the electrode to represent the characteristic deposition rate of a system.

The system pressure dependences of the deposition rate onto the electrode surface in DC, 40 kHz, and 13.56 MHz discharges are shown in Figure 8.18. As anticipated from the deposition rate equation, given in Eq. (8.7), the deposition rate of DC cathodic polymerization is linearly proportional to the system pressure. The deposition rate in 40 kHz discharge was found to be pressure dependent also, but that in 13.56 MHz was found to be independent of system pressure. The deposition rate in the 40 kHz discharge is one half of that in the DC discharge, and the slope of pressure dependence is also roughly one half of that obtained from DC discharge. These findings indicate that the material formation in the molecular dissociation glow takes

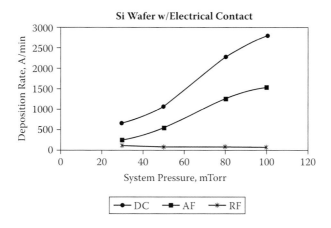

FIGURE 8.18

The system pressure dependence of deposition rate of TMS on Si wafer with electrical contact to the powered electrode in DC (cathode), 40 kHz, and 13.56 MHz plasma polymerization processes; discharge conditions are 1 sccm TMS, 5 W power input.

place on the electrode in a 40 kHz discharge. As the frequency increases to 13.56 MHz, the electrode does not act as the cathode, such as in DC or 40 kHz discharges, and the material formation in the diffuse luminous gas phase then governs the deposition onto the electrode. Because the dissociation glow exists very close to an electrode, the deposition rate onto an electrode is higher than that on a floating substrate; however, the deposition on an electrode of RF discharge is that of glow discharge polymerization that occurs in the diffused luminous gas phase, and the rate is lower than those in DC and 40 kHz discharges.

8.4.3 Deposition on Floating Substrate in Luminous Gas Phase

TMS deposition rate profiles on floating substrate in DC, 40 kHz, and 13.56 MHz discharges are shown in Figure 8.19, in which the scale of the deposition rate is 0 to 300, whereas that in Figure 8.17 is 0 to 1,400. The deposition rate on floating substrate is smaller than that on an electrode in all cases, but uniform deposition occurs on the substrate. In DC discharge, there is a significant difference in the deposition rate on the surface facing the cathode and that on the surface facing the anode. This difference is proof that the cathode glow (dissociation glow) prevails in DC discharge, and the reactive species created in the dissociation glow (at the cathode surface) is diffuse through the gas phase. This concept of diffusive transport of reactive species created in the cathode glow (dissociation glow) is confirmed in the following section of the role of anode in DC discharge. Hence the distance from the cathode is a factor that controls the deposition onto substrate placed in the gas phase.

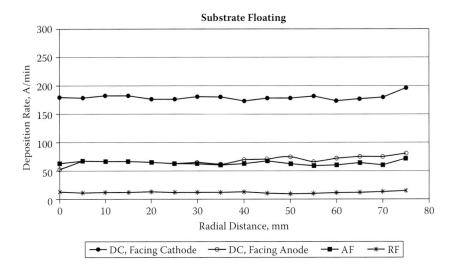

FIGURE 8.19
Deposition rate profiles of TMS in DC, 40 kHz, and 13.56 MHz LCVD on floating substrate; 1 sccm TMS, 50 mtorr system pressure, 5 W power input.

In order to see the influence of electrical contact, some silicon wafers were electrically insulated from the substrate plate used as the cathode by placing a thin slide cover glass between the silicon wafer and the substrate. The influence of the electrical contact on deposition rate onto the electrode and onto the floating substrate is shown in Figures 8.20, 8.21, and 8.22 as a function of system pressure. In the right side of the figures, the influence of the same factors on the refractive index is shown. The scale of the deposition rate

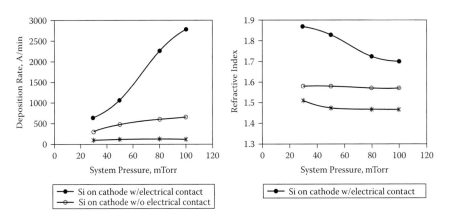

FIGURE 8.20
The system pressure dependence of deposition rate (left) and refractive index (right) of TMS on Si wafer with electrical contact and without electrical contact to powered electrode or substrate floating in gas phase for DC LCVD; 1 sccm TMS, 5 W power input.

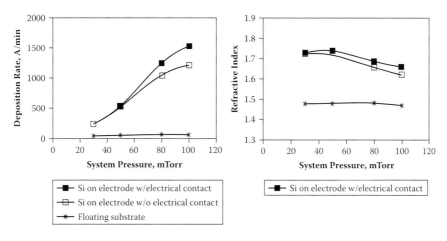

FIGURE 8.21

The system pressure dependence of deposition rate (left) and refractive index (right) of TMS on Si wafer with electrical contact and without electrical contact to powered electrode or substrate floating in gas phase for 40 kHz LCVD; 1 sccm TMS, 5 W power input).

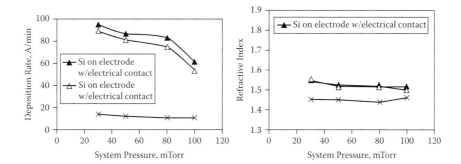

FIGURE 8.22

The system pressure dependence of deposition rate (left) and refractive index (right) of TMS on Si wafer with electrical contact and without electrical contact to powered electrode or floating substrate in gas phase for 13.56 MHz LCVD; 1 sccm TMS, 5 W power input.

axis is different for each case in order to show the system pressure dependence clearly.

In a DC discharge (Figure 8.20), without electrical contact, the deposition rate differs significantly from that for *cathodic polymerization* (with electrical contact), and the pressure dependence is marginal. If we consider the substrate without electrical contact on a cathode surface as the floating substrate at distance 0 from the cathode, the deposition rate on the 0 distance floating substrate should be proportional to the deposition rate on the cathode, which explains the marginal pressure dependence. The deposition onto a floating substrate can be characterized as typical plasma polymerization (material formation in the diffused luminous gas phase). A major difference in refractive index is seen between the electrode and nonelectrode use of the substrates.

In a 40 kHz discharge (Figure 8.21), the deposition onto the surface of the electrode, regardless of electrical conductivity or contact, is significantly different from deposition onto a floating substrate. The cathodic aspect of the electrode is less (one half of DC discharge), but because of this, the overall cathodic aspects of polymerization extend beyond the surface of the electrode, yielding cathodic plasma polymer on an electrically insulated substrate placed on the electrode. Thus, the features of cathodic polymerization dominate in the vicinity of the electrode regardless of electrical contact.

In 13.56 MHz discharge (Figure 8.22), the deposition on the substrate placed on an electrode, regardless of electrical conductivity or contact, is also appreciably different from that on the floating substrate, although the magnitude of the difference is much smaller than that found in DC or 40 kHz discharges. However, the deposition onto the electrode has no feature of cathodic polymerization. In 13.56 MHz discharge, the deposition of materials is primarily by plasma polymerization, and the deposition rate is given by Eq. (8.3). The higher deposition rate on the electrode could be explained by the effect of the molecular dissociation glow being close to the electrode surface. In RF discharge, the site of activation has shifted away from the electrode surface, and the features of clearly identifiable cathodic polymerization are diminished (i.e., the pressure dependence of deposition rate on an electrode is slightly negative in RF discharge).

The material formation in LCVD is caused mainly by the dissociation glow (DG), and the ionization glow (IG) consists mainly of non-polymer-forming species. In DC discharge, the material formed in the cathode glow deposits nearly exclusively on the cathode surface due to the adherence of DG to the cathode, but some could deposit on surfaces existing in the reactor. The situation with the material formed in the negative glow is the same (i.e., it could deposit on the cathode, the anode, and surfaces placed in the reactor). With electrically floating substrates, the deposition rates, as well as the refractive indices, are nearly the same for DC and 40 kHz glow discharges. Under the set of conditions employed, 13.5 MHz discharge yielded the lower deposition rate, but the refractive index was found to be nearly the same as those samples formed in DC and 40 kHz discharges. This implies that the material formation in the diffused luminous gas phase of DC and that in 40 kHz and 13.56 MHz are essentially the same.

8.4.4 Role of Anode in DC Cathodic Polymerization

In DC cathodic polymerization conducted in a bell jar reactor, the cathode (substrate) is positioned in the middle between the two anodes. In such an electrode arrangement, the distance between the cathode and the anode is expected to have some effects on the deposition rate and deposition profile with respect to those without anode assembly. Figures 8.23 and 8.24 show

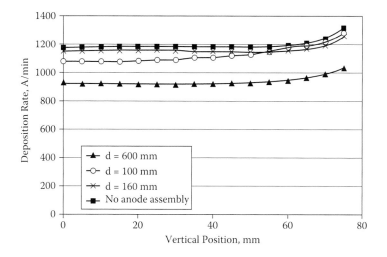

FIGURE 8.23
The influence of electrode distance on the deposition rate on cathode surface in DC cathodic polymerization; 1 sccm TMS, 50 mTorr, DC 5 W; *d* is the distance between two anodes, *d*/2 is the distance between the cathode and an anode.

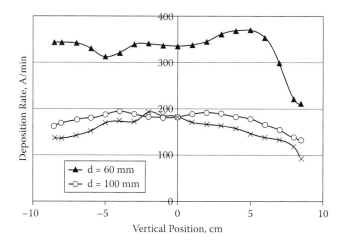

FIGURE 8.24
The influence of electrode distance on the deposition rate on an anode surface in DC cathodic polymerization; 1 sccm TMS, 50 mtorr, DC 5 W; *d* is the distance between two anodes, *d*/2 is the distance between the cathode and an anode.

the influence of the distance between two anodes (one half of which is the cathode–anode distance) on TMS deposition rate on a cathode (i.e., substrate) and an anode, respectively.

From Figure 8.23, it can be seen that with the increase of anode spacing from 60 mm to 160 mm, the deposition rate on a cathode (substrate) showed an increasing trend. The deposition on a cathode (substrate) surface seemed to reach the maximum when the anodes were removed from the plasma system (i.e., no anode assembly was present and the grounded reactor wall functioned as an anode). In contrast, it is noted that, from Figure 8.24, the deposition on an anode surface decreased with the increase of anode spacing. These results clearly indicated that the too close anode spacing could not only decline the preferred plasma polymer deposition on substrate (cathode) but also induced more undesired deposition on the anode surface. In other words, DC cathodic polymerization without using an anode assembly seems to be a more efficient and realistic method for practical applications.

As seen in Figure 8.23, DC cathodic polymerization of TMS without using an anode assembly gave rise to higher deposition than that with an anode assembly. Therefore, the nature of anode and the role it plays in the DC cathodic polymerization process needs to be further clarified. In DC discharge, as noted from Figures 8.23 and 8.24, the much smaller amount of deposition occurs onto the anode surface than occurs onto the cathode. Figures 8.25 and 8.26 show the effects of the floating panels positioned between anode and cathode on the deposition rate on an anode surface and a cathode surface in DC discharge. The floating panels did not affect the plasma deposition on cathode. They acted just as a surface cover (2 cm away) of the anode and showed a similar deposition rate to that on anode. The deposition on an anode surface declined at the edge of the floating panel, and no deposition was detected in the center of the anode.

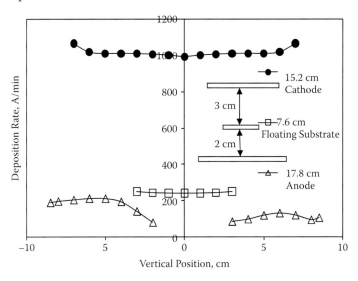

FIGURE 8.25
The effect of electrically floating substrate panel positioned in front of an anode on deposition on an anode surface and on an anode surface in DC cathodic polymerization; 1 sccm TMS, 50 mTorr.

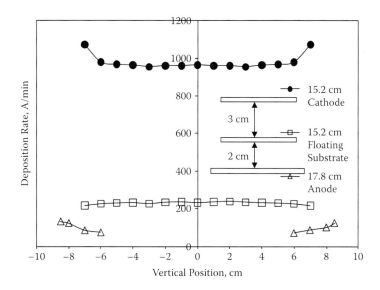

FIGURE 8.26
The effect of electrically floating larger substrate under the same conditions of Figure 8.25.

These results indicate that the anode is a passive surface as far as the plasma polymerization is concerned, and the same deposition occurs onto the floating substrate placed in between the cathode and anode. This also means that the anode, as a passive surface, collects polymerizing species created by the cathode glow (dissociation glow). When the passive surfaces are eliminated, the share of deposition onto the cathode increases slightly as seen in Figure 8.23. These data clearly show the fundamental mechanism of plasma polymerization that reactive species are mainly created in the dissociation glow, and the creation in negative glow, if any, is marginal, and negative glow plays the role of sustaining the energy level of photon-emitting species as demonstrated by the magnetic field shaping of negative glow in Chapter 6.

8.4.5 DC Plasma Polymerization in a Closed System

In flow system LCVD, the system pressure is continuously adjusted by controlling the opening of a throttle valve connected to the pumping system. Because of fragmentation of the original monomer in the plasma state, the composition of the gas phase changes on the inception of the plasma. The increase in the total number of gas molecules is compensated for by the increased pumping rate in a flow system, and a steady-state flow of a consistent composition of gas phase is established at a predetermined system pressure.

In closed-system LCVD, a fixed amount of monomer molecules are contained in the reactor, and glow discharge is initiated. The pressure in such a system (in a given volume) is proportional to the total number of gas phase molecules. The fragmentation of monomer molecules as well as the ablation of gaseous species from the deposited material will increase the pressure, and deposition will decrease the system pressure. Thus, the change in system pressure with plasma polymerization time will indicate the change in the overall balance between the plasma fragmentation/ablation and the plasma polymer deposition.

In DC glow discharge of TMS in a closed reactor, the system pressure increases when the discharge is initiated. The change of system pressure is shown in Figure 5.9 (Chapter 5). The system pressure continuously increases while the glow discharge is on, but it remains at a constant value as soon as the glow discharge is turned off. This indicates that the total number of gas phase species increases with time in spite of the deposition of plasma polymer of TMS. A residual gas analyzer (RGA) characterized the gas phase composition of TMS plasma in the closed reactor system, and results are shown in Figure 8.27. It can be seen that the main gas phase species from the fragmentation of TMS monomers are hydrogen molecules and carbon-containing or silicon-containing molecular segments. The white unfilled bars indicate the dissociation pattern of TMS in RGA (without plasma).

In the early stage of the discharge, the first 15 seconds, silicon-containing species were the dominant species in the gas phase system. Because of the fast deposition characteristics of silicon species, the silicon-containing species disappeared very quickly in the early stage of glow discharge. In

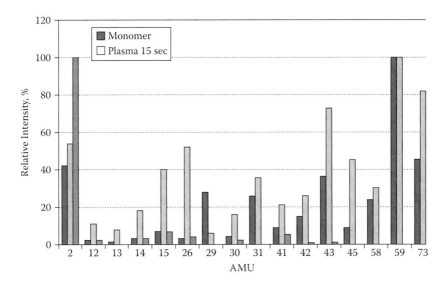

FIGURE 8.27
The significant gas phase component observed in a closed-system TMS DC discharge. The monomer is not exposed to plasma and represents the cracking pattern of TMS in the RGA.

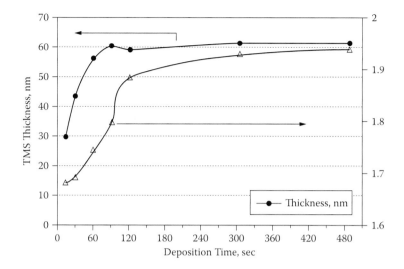

FIGURE 8.28
Change in thickness and refractive index of TMS plasma polymer coating with discharge time in a closed reactor system.

the final stage, after 120 seconds, both the silicon-containing and carbon-containing species have been consumed by the LCVD deposition. In the gas phase, only hydrogen is left in the plasma system, and no further deposition occurs. Therefore, it is anticipated that there will be no further thickness growth of the TMS plasma coatings after 120 seconds.

Figure 8.28 shows the time dependence of the thickness and refractive index of TMS plasma coatings on the plasma polymerization time in a closed reactor system. The coating thickness increases very fast in the first 90 seconds. After 90 seconds, the TMS coating thickness stops growing with additional deposition time simply because no TMS is left in the system. On the other hand, the refractive index of the TMS coating continues to increase in a remarkable extent with the deposition time. This increase obviously results from continuing interaction with luminous gas of hydrogen.

The characteristic plasma deposition rate of Si-containing organic compounds is nearly six times greater than that of hydrocarbons [14], and it is anticipated that Si-containing moieties would deposit faster than C-based moieties leading to Si-rich depositions from TMS, which contains one Si and three C in the original molecule. This difference in the characteristic deposition rates would be amplified if the plasma polymerization is carried out in a closed system, because the gas phase composition with respect to Si and C changes rapidly and continuously as plasma polymerization proceeds, and the composition of the deposition changes accordingly.

According to this scheme of closed-system plasma polymerization of TMS, it is anticipated that the atomic composition of the plasma polymer should continuously change with the plasma polymerization time. The comparison of XPS depth profile-generated C/Si ratios for plasma polymers deposited in both a flow-system reactor and in a closed-system reactor is shown in Figure 7.5 (Chapter 7). The TMS plasma coating prepared in a closed system shows gradual composition changes from lower carbon content (C/Si ratio of ~2.0) at the interface with the substrate to carbon rich (C/Si ratio of ~3.8) at the surface as deposition proceeds. In contrast, the coating obtained in the flow system had a uniform composition (C/Si ratio of ~1.5) throughout the film. These numerical values, of course, depend on operational parameters, such as the size of the reactor, the size of the cathode, the current density, and the amount of TMS introduced into the reactor. Nevertheless, these results clearly show that closed-system plasma polymerization of TMS indeed produces a film with graded composition (i.e., with increasing carbon content from the film/metal interface to the top surface of the nanofilm).

Considering the fact that the refractive index continues to increase after most of the polymerizable species are exhausted in the gas phase, DC LCVD of TMS in a closed system contains the aspect of luminous chemical vapor treatment (LCVT) of deposited plasma polymer coating by hydrogen luminous gas phase. In the later stage of closed-system LCVD, oligomeric moieties loosely attached to a three-dimensional network are converted to a more stable form, and significantly improved corrosion protection characteristics compared to the counterpart in flow-system polymerization of TMS were found. Thus the merit of closed-system cathodic polymerization is well established, which creates a film strongly adhering to a metallic substrate that changes gradually to the top layer with better adhesion characteristics than a conventional corrosion protection coating.

References

1. Matsuzawa, Y. and L. Winterton, Presentation at the International Round Table on Plasma Interface Engineering, Surface Science and Plasma Technology, University of Missouri, Columbia, 2002.
2. Yasuda, H. and T. Hirotsu, *Journal of Polymer Science: Polymer Chemistry Edition*, 16, 743, 1978.
3. Yasuda, H., *Plasma Polymerization*, Academic Press, Orlando, FL, 1985, 277–333.
4. Miyama, M. and H. Yasuda, *Journal of Applied Polymer Science*, 70, 237, 1998.
5. Yasuda, H. and C. R. Wang, *Journal of Polymer Science: Polymer Chemistry Edition*, 23, 87, 1985.
6. Yasuda, H. and Q. Yu, *Plasma Chemistry and Plasma Processing*, 24, 325–351, 2004.
7. Yu, Q. S., C. Huang, and H. K. Yasuda, *Journal of Polymer Science. Part A, Polymer Chemistry*, 42, 1042–1062, 2004.

8. Yasuda, H. and Q. Yu, *Journal of Vacuum Science and Technology. A, Vacuum, Surfaces, and Film*, 22(3), 472–476, 2004.
9. Yasuda, H., *Luminous Chemical Vapor Deposition and Interface Engineering*, CRC Press, Boca Raton, FL, 2004.
10. Yasuda, H., *Plasma Processes and Polymers*, 2, 293–304, 2005.
11. Yasuda, H. K. and Q. S. Yu, *Journal of Vacuum Science and Technology. A, Vacuum, Surfaces, and Film*, 19(3), 773, 2001.
12. Yu, Q. and H. Yasuda, *Plasma and Polymers*, 7, 41, 2002.
13. Yu, Q., C. Moffitt, D. Wieliczka, and H. Yasuda, *Journal of Vacuum Science and Technology. A, Vacuum, Surfaces, and Film*, 19(5), 2163, 2001.
14. Gazicki, M. and H. Yasuda, *Applied Polymer Symposia*, 18, 35, 1984.

9

Magneto-Luminous Chemical Vapor Deposition

9.1 Domain of Magneto-Luminous Chemical Vapor Deposition (MLCVD)

Plasma polymerization in general is not an inexpensive and easy process, as some researchers tend to promote the image of the technology when it comes to the stage of applying it in an industrial-scale operation. It is a simple and relatively inexpensive process in the laboratory to start with. However, some factors, which are not considered as important, or are neglected in laboratory experiments, often show up as factors that cause problems in industrial scale operations; i.e., the scale up of a process, which showed promising results in laboratory, to an industrial scale operation is not an easy undertaking. In this particular aspect, magneto-luminous chemical vapor deposition (MLCVD), or magnetron plasma polymerization, is unique. It could be the only plasma polymerization method that has very impressive credentials in the following respects:

1. Magneto-plasma polymerization can be operated for a long period of time in a continuous mode without being hampered by excessive deposition elsewhere beyond the substrate surface, on which the deposition of plasma polymer is aimed for.

2. The above feature has been proven by over 10 years of large-scale industrial operation in continuous mode for one month between short preventive maintenance breaks.

3. It is ideally suited for creation of an "imperturbable surface state" in the range of 20 nm to 50 nm that is implanted into the adequately prepared surface state of the substrate materials, metallic or polymeric, by the front-end green process.

4. The product produced by the method cannot be prepared by any other method, plasma or conventional chemical or physical processes.

Plasma polymerization started as a low-pressure electrical discharge process, of which the major cost factor was the cost of the vacuum reactor. It is interesting to note that the pioneering group of "plasma polymerization" in the early 1960s developed later an excellent hybrid process of plasma grafting in the mid-1970s, in which Ar glow discharge was used to create free radicals on polyester fabrics and then was transferred to the treated fabrics in vacuum, in a continuous manner, to the second chamber, in which vapor of acrylic acid was introduced to graft polymerize acrylic acid on the surface of the fabrics by free radicals formed by Ar plasma in the first chamber. Namely, they developed a commercial-scale hybrid process of Ar plasma surface treatment of fabrics and conventional free-radical polymerization of acrylic acid as a dry process. This process produced excellent surface-modified fabrics that had durable soil resistance and high liquid water-wicking characteristics at a low processing cost, because the process used a steam aspirator instead of a vacuum pump to reduce the pressure in the reactor to the operating pressure. It is obvious that the cost of the vacuum system was a major concern in the effort to develop an industrial-scale plasma reaction system.

This example casts a few crucial questions dealing with low-pressure plasma processes:

1. Is plasma polymerization of acrylic acid best for attaining characteristics of the final product?
2. Is the hybrid process better than plasma polymerization with respect to the process to attaining the final product, though there is advantage in the cost and the ease of operation?
3. What are the specific objectives that require plasma processes?
4. Can a conventional wet chemistry process produce the same effect on the surface of fabrics? In other words, is the plasma process absolutely necessary or does it have decisive advantages over other conventional chemical processes?

Although the hybrid plasma surface modification was a technological success, the process did not succeed as a successful commercial operation, simply because the company that purchased the process, a textile treatment company, could not find a decisive market for such an excellent product within a short pre-perceived market development time. Quite coincidentally, a textile company developed more or less the same product by conventional wet chemical process at around the same time and eventually found the decisive commercial markets for their products. However, it is important to note that it took a few years of soul-searching efforts by textile experts to find the market for such a unique product.

Reviewing characteristics and performance of both products, the author reached the judgment that the product produced by the plasma hybrid process might have had a slight edge in the durability of surface characteristics, but the product prepared by the conventional wet chemistry was certainly good enough for the market they developed. However, if the green aspects of production processing were taken into account, judgment could have shifted in either direction.

The above episodes, on developing seemingly the same surface modified fabrics by (a) plasma process and (b) conventional wet-chemistry, point out some key factors that are important in developing plasma polymerization processes for commercial applications.

1. Technical interest should not be the main driving force for the technology development. Both processes mentioned above were evidently developed not fulfilling market demand, but to produce unique properties that were not available with polyester fabrics.

2. Successful technology development generally follows the push–pull principle, in which scientists and engineers are only on the push side. Whether push or pull comes first is not the issue.

3. Developers of technology are usually overconfident on what they achieved without enough knowledge of requirements on the pull side.

As far as low-pressure plasma polymerization is concerned, it is absolutely necessary to identify the decisive advantage of the process under investigation, which cannot be obtained by other means. It is also imperative to examine whether the process could be operated in a continuous mode in order to bring the equipment depreciation cost to a small fraction of the cost of the final product, as emphasized in Chapter 3. In this particular requirement, MLCVD has unequivocal advantages and a proven track record.

The equipment cost cited in Figure 3.3 (Chapter 3) could provide an example to point out how important the capability to operate plasma polymerization in a continuous manner and in a green process is. If the plasma polymerization reactor, which cost $3 million, can produce 30,000 products in 10 years, the depreciation cost per product, which should be included in the cost of product, is $100. This cost decreases with the number of products in 10 years: $1.00 for 3 million, and $0.01 for 300 million. The acceptable value depends entirely on the add-on value created by the process. In the case of the coating of contact lenses, the value comes down to 1 cent per lens, which means that the plasma polymerization process must produce 300 million coated contact lenses in a 10-year period. It also means the plasma polymerization coating reactor can be operated continuously for at least 30 days consistently without break, and the operation can be continued for a 10-year period without any major breakdowns and frequent and lengthy maintenance breaks. Such a stringent requirement cannot be fulfilled if plasma polymerization cannot be operated in a continuous manner. The MLCVD process has the very unique track record in over 10 years of fulfilling such a stringent requirement with over 10 reactors in operation.

Factors that make MLCVD uniquely different from other low-pressure plasma polymerization are as follows:

1. Magnetic field enhancement causes a completely different gas phase breakdown mechanism, which enables us to operate in the unique domain of the gas phase breakdown phase diagram (Figure 6.18, Chapter 6) than domain [III] of ordinary gas phase breakdown phase diagram (Figure 5.32) by LCVD (without magnetic field).

2. The power source is 15 kHz symmetric power that creates the dissociation glow in two symmetrical toroidal glows, which lift the cathode glow from the cathode surface to the gas phase and keep the toroidal glow surface of the cathode open without deposition of plasma polymer. Without this feature, the long-time continuous operation of plasma polymerization is practically impossible.

3. The low pressure of operation of the magneto-luminous gas phase enables us to operate MLCVD in the low-pressure domain that cannot be reached by other modes of discharge. The low-pressure operation yields uniform deposition of amorphous carbon nanofilm with extremely good adhesion onto smooth substrate surfaces of various kinds of materials adequately prepared by system approach interface engineering.

4. The types of gas used are the simplest organic molecules, such as CH_4, which has been extensively explored, and $HSi(CH_3)_3$, HF_2C-CF_2H, and so forth, which have not been well explored at the time of writing this book. These molecules do not contain easily plasma-polymerizable functional groups such as double bond, triple bond, cyclic structure, and polymerize through only mono-radical cycle 1 of the repeating step growth polymerization (RSGP) mechanism (see Figure 7.1, Chapter 7). This selection of simple monomers that utilize only mono-radical cycle 1 of the RSGP mechanism virtually eliminates the possibility of change of the polymerization/deposition mechanism with operation time due to the contribution of functional groups, which operate in cycle 2 paths of the RSGP mechanism.

5. This selection of simple monomers also reduces the amount of trapped or residual free radicals in the applied nanofilms. The trapped free radicals in a tight network, per se, are not so serious a problem in the relatively short service life of the coated product, because they are embedded in a very tight network (dangling bonds), and O_2 molecules cannot access them. Nevertheless, the lesser amount of trapped free radicals is preferred for sustainable function of the coated products, particularly in biomedical applications.

6. Concerning moving substrates in the luminous gas phase, the substrates move continuously by a rotating sample holder that transports substrates in and out of the luminous gas phase. Without

relative motion between substrate surface and luminous gas phase, uneven distribution of mass and properties of nanofilm, in undesirable levels, are inevitable.

7. MLCVD can be operated either in batch operation mode for small objects or for a small number of products, or in large-scale continuous operation as described in Chapter 3. The sustainability and the reproducibility of plasma polymerization are the key issues.

8. With MLCVD, the scale up of laboratory experiments to a larger scale operation is nearly guaranteed, because it is only necessary to modify the substrates transport mode depending on the product volume. In contrast to this situation, the scale up of a plasma polymerization carried out in laboratory often means the startup of a new game, which is often a large scale new experiment, not a large scale production of what is made in the laboratory scale experiments.

9. The distribution patterns within a substrate (silicon wafer) confirm that substrates could be mounted on both sides of substrate holding wheel, which rotates and gives in-and-out motion of substrate in a bell-jar type reactor used in the laboratory. It also confirms, through contact lens coating, that if a sheet of film was mounted so that two sides of film are exposed to the luminous gas phase, both sides could be coated by a single operation. Sheet of film or fabrics could be pulled in the middle of two electrodes continuously, removing the sample holding wheel, to coat both sides or two sheets back on back to coat one side; these modification involve only the substrate feeding system. Consequently, scale up of operation from laboratory experiment to industrial large scale production is a straight forward undertaking, because no fundamental change of MLCVD process is involved.

9.2 Toroidal Glow Surface without Deposition

As the consequence of having a magnetic field superimposed on the cathode surface, the glow that develops with MLCVD is different from that which occurs without a magnetic field as described in Chapter 6. In the MLCVD process, the toroidal glow is the only glow that develops near the cathode as depicted in Figure 6.3 (Chapter 6). The surface of the cathode under toroidal glow, which is in the gas phase, has no deposition of plasma polymer and remains open as depicted in Figure 9.1. This is in strong contrast to the glow that develops without a magnetic field (i.e., the dissociation glow covers the entire cathode surface, and the main deposition of plasma polymer occurs on the electrode surface). Because the toroidal glow surface is open without deposition, the MLCVD can be operated without being hampered by the deposition onto the electrode surface and enables us to run the process for a long time. This is a

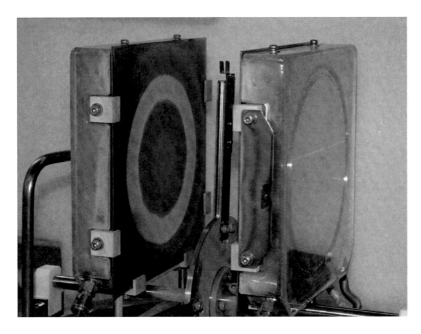

FIGURE 9.1
Deposition pattern on the magnetron cathode. No deposition of plasma polymer on the toroidal glow surface.

unique feature of the MLCVD process. Without this feature, plasma polymerization characteristics change with the operation time, and consequently, the deposition characteristics also change with time (i.e., the continuous operation of LCVD or plasma polymerization is practically impossible).

In certain operations of MLCVD with CH_4, a certain amount of oxygen or air is added in the feed gas. The main purpose of adding oxygen or air is not for making the surface of coated film hydrophilic but for ensuring the clean opening of the toroidal glow surface without deposition in long-time operation of the process. As described in Chapter 7, oxygen does not participate in plasma polymerization (iN/Out rule), and the surface oxidation mainly occurs by the reaction of oxygen in ambient air with free radicals left in the plasma polymer film when the coated substrate is taken out of the reactor.

The main effect is that the oxidative etching of materials deposits, if it occurs, on the toroidal glow surface of the cathode and keeps the toroidal glow surface clean without deposition. It is interesting to note that the oxygen cleaning of a plasma reactor does not work with magnetron, simply because oxygen plasma only clean the toroidal glow surface of the electrode. This cleaning of toroidal glow surface works simultaneously with a MLCVD of CH_4 operation when O_2 is added in the feeding gas.

Another hidden advantage of adding O_2 or air is the reduction of deposition rate, which might sound counterproductive; however, one should keep in mind that the most important factor of nanofilm coating is the quality of the product, not how fast one could produce. The addition of O_2, of course, reduces the deposition rate by the same mechanism described for O_2 plasma cleaning of the toroidal glow surface, but the reduction of the deposition rate works in favor of upgrading the quality of the nanofilm. The reduction of the number of reactive species in the luminous gas phase shortens the kinetic path length of the material formation, which yields better adhesion of smaller grains, beyond the extent obtainable by only low-pressure operation.

9.3 Confined Luminous Gas Phase in Low Pressure

In general, the volume of luminous gas phase within a reactor increases as the system pressure decreases, and the area in which deposition of LCVD products occurs increases far beyond the energy input area. This situation could be visualized by the figures of glows shown in Figure 9.2. This picture

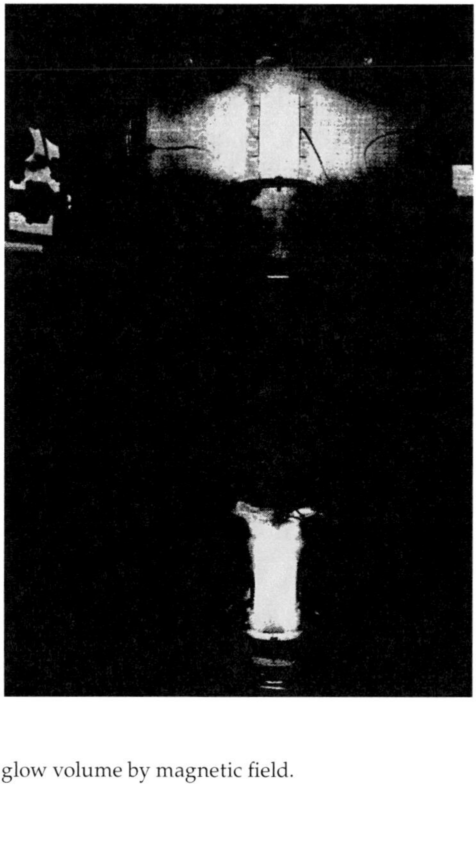

Confining of Glow Volume

Without Magnet

With Magnets

FIGURE 9.2
Confining of glow volume by magnetic field.

was taken using a tandem reactor that had two identical plasma polymerization reactors equipped with MLCVD capability stacked vertically. The moving substrate continuously travels vertically from the bottom reactor to the top reactor and to the winding system above. The operational conditions of each reactor could be set depending on the specific need of a two-stage coating process. In order to see the confinement effect of the magnetic field, the magnets in the top reactor were removed, and Ar discharges were created under otherwise identical operation conditions. What we see are overall volumes of negative glow resulting from the identical glow discharge operation, except with and without magnetic fields superimposed on electrodes.

It is very clear that the glow volume is confined by the magnetic field, and the intensity of glow is higher in the bottom reactor. With deposition gas, the area of the reactor wall in contact with glow collects the deposition of plasma polymer, which means wall contamination increases with the expansion of glow volume, and reduces the yield of deposition onto substrate that passes through the middle of interelectrode volume. Without magnetic field confinement of glow volume, the effort to take advantage of the low-pressure operation of LCVD backfires in increasing the reactor-wall contamination and reducing the yield of deposition on substrate. The benefits of MLCVD over LCVD are clear—the increase of yield of deposition onto substrate and the reduction of wall contamination.

9.4 Polymer Formation and Deposition in Low Pressure

9.4.1 Deposition Kinetics in Domain M

Luminous chemical vapor deposition, or plasma polymerization, by 15 kHz (in general 10 to 40 kHz) power input is essentially alternating direct current (DC) discharge plasma polymerization. In DC, discharge of deposition gas is controlled by the current density in the reactor as shown in Figure 8.16 (Chapter 8). The energy input parameter (based on the luminous gas phase) does not control the deposition of material onto the cathode surface, and the flow rate of monomer does not influence the film thickness growth rate. The film thickness growth rate is dependent on the mass concentration of monomer (c_M) in the cathode region rather than the mass input rate (FM).

In MLCVD, the power input is essentially the same as with DC discharge, but the domain in which plasma polymerization occurs is totally different, as described in Chapter 7. How the difference of the domain of gas phase breakdown influences the deposition kinetics of MLCVD could be seen by examining the deposition rate, or thickness growth rate, dependence on operational parameters, as shown in the following figures.

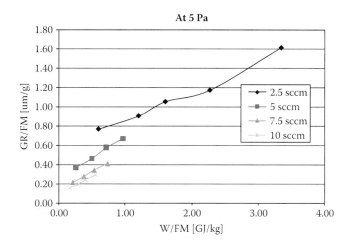

FIGURE 9.3
Dependence of film growth rate *GR/FM* on power input parameter *W/FM*.

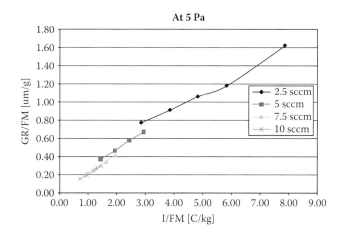

FIGURE 9.4
Dependence of film growth rate *GR/FM* on power input parameter *I/FM*.

The dependence of film growth rate per mass flow rate of deposition gas, *GR/FM*, on energy input parameter, *W/FM*, is shown in Figure 9.3. In radio frequency (RF) discharge, all data points would be on a single line. Although the general trend that the deposition rate increases with *W/FM* is evident, the plots indicate that the deposition kinetics in MLCVD is different from that in RF discharge. Figure 9.4 depicts the dependence of *GR/FM* on *I/FM*, which shows less deviation due to the flow rate but still shows unsatisfactory influence of flow rate.

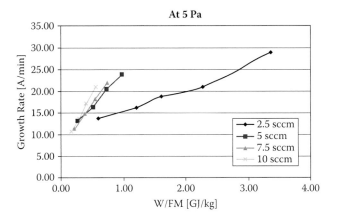

FIGURE 9.5
Plots of film growth rate against *W/FM*.

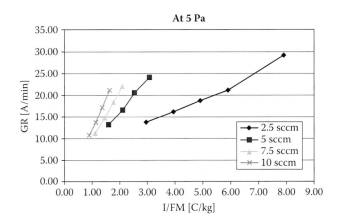

FIGURE 9.6
Plots of film growth rate against *I/FM*.

These figures also indicate that the mass near the cathode surface rather than flow rate into the reactor controls the deposition rate as mentioned at the beginning of this section. Figures 9.5 and 9.6 depict plots of film growth rate plotted against *W/FM* and *I/FM*, respectively. Both figures indicate that (*W/FM*) and (*I/FM*) are the operational parameters that control the film growth rate in the MLCVD process. Plots of film growth rate against discharge current, depicted in Figure 9.7, show the direct proportionality of growth rate and the discharge current, which also shows the slight decline of the growth rate at higher flow rates. This trend might be reflecting the decrease of resident time of gas molecules in the luminous gas phase. In any case, the film thickness increases linearly with reaction times at a fixed

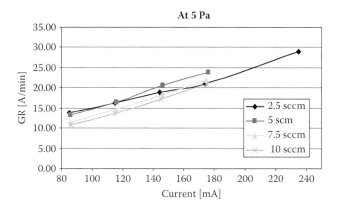

FIGURE 9.7
Plots of film growth rate against discharge current.

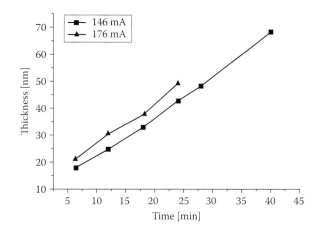

FIGURE 9.8
Thickness growth with operation time at a fixed current.

current as depicted in Figure 9.8, which enables us to set a standard procedure to control film thickness under preset operation conditions.

The fact that the deposition rate by MLCVD is controlled by the discharge current is an extremely important feature, because the main feature of the breakdown process under the influence of a magnetic field is the extremely high current at low voltage. In other words, the polymer deposition by MLCVD at very low pressure proceeds much faster than one might conceive by conventional wisdom of chemical reaction in low pressure. Without recognition of the current dependent deposition rate and the exceptionally high current in low pressure (operation in domain [M] of gas phase breakdown phase diagram), what is happening in the MLCVD process cannot be fully comprehended.

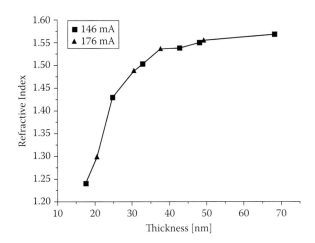

FIGURE 9.9
Dependence of refractive index of film on the film thickness.

Reflecting the repeating electron impact excitation and coupling of excited species according to the RSGP mechanism of film formation in the luminous gas phase, the refractive index of film is dependent on the thickness of deposited film as depicted in Figure 9.9. The discharge current seems to have little influence on the refractive index change, which indicates the change is due to the deposition kinetics rather than the excitation kinetics. This effect can be clearly seen also in the closed-system DC cathodic polymerization of TMS, depicted in Figure 8.28 (Chapter 8) (i.e., after film thickness increase stopped due to extinction of monomer gas in the system, further exposure of the deposited film increases the refractive index).

9.4.2 Pressure Dependence of Deposition Rate

The dependence of deposition rate on system pressure in audio frequency (AF) discharge depends on electrical contact to the substrate and does not depend on electrically floating substrate. In MLCVD, the deposition always occurs onto electrically floating substrate, but the deposition rate depends on the system pressure as depicted in Figure 9.10, because the discharge current in domain [M] depends on the system pressure. Because the refractive index of MLCVD nanofilm is dependent on the film thickness, the influence of system pressure on the refractive index follows similar dependence as depicted in Figure 9.11.

The numbers representing three lines are the distances from an electrode to the substrate, which is one half of electrode-to-electrode distance. The density of gas increases with pressure, and the deposition rate usually increases with pressure. The negative dependence of the deposition rate on system pressure is one indication that MLCVD is not one mode of LCVD or

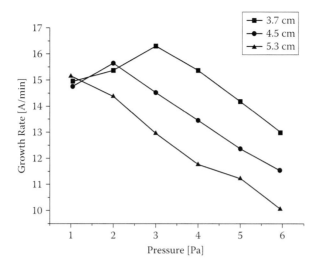

FIGURE 9.10
The dependence of film growth rate on system pressure of MLCVD operation. The numbers shown are distances from electrode to substrate (one half the distance of electrode to electrode).

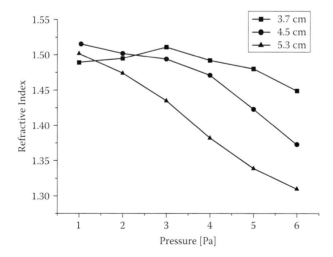

FIGURE 9.11
The dependence of refractive index of MLCVD nanofilm. The decrease of the refractive index is due to the decrease of thickness at higher pressure.

plasma polymerization, but a unique deposition process in [M] domain gas phase. Although more fundamental investigation of material formation in the [M] domain of a broken-down gas phase is needed to explain the phenomenon, it is likely that the decline of deposition rate with increasing pressure is caused by the decrease of discharge current in high pressure in the

[M] domain, because the deposition rate is controlled by the discharge current in MLCVD.

The pressure dependence described in this section confirms the great advantage of MLCVD that operates in low pressure, but at the same time provides warning that MLCVD is not meant to be operated in higher pressure.

9.4.3 Small Grain Size and Uniform Smooth Surface

The low pressure and high current at low-voltage operation of MLCVD yields unique advantageous characteristics of the product nanofilm. The high current yields a much higher deposition rate in the lower pressure than ordinary plasma polymerization (by RF discharge), due to the unique current-driven plasma polymerization mechanism of MLCVD. Without this feature, plasma polymerization in such a low pressure would have been impractical due to a low production rate of the nanofilm.

As briefly pointed out in Chapter 7, the size of the depositing entity has very important effects on many crucially important characteristics of the nanofilm obtained. Gas molecules contained in a vessel collide with gases and also with the surface that contains them. The ratio of (gas–gas collisions)/(gas–surface collisions) in a vessel depends on the gas pressure (mean free-path) and increases with pressure. In the luminous gas phase, collision between reactive species increases the size of gaseous species, which leads to the deposition of larger species at higher pressure of reaction vessel, which could be visualized as the increase of "grain size" of the deposited film. This situation is schematically depicted in Figure 9.12. It can be easily conceived that factors that increase the kinetic path length in the gas phase would yield deposition of larger grain (less tight) with

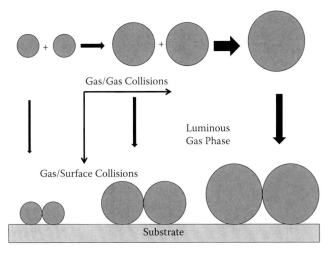

FIGURE 9.12
Influence of gas–gas collision and gas–surface collision.

lesser contact area with substrate (less adhesion). The addition of O_2 to CH_4 works in the same way by reducing the kinetic path length as mentioned earlier.

9.4.4 Strong Adhesion to Substrate Surface

It is also clear, from Figure 9.12, that the ratio of mass/contact area between the surface and deposited particle increases with the system pressure, which means that the adhesive force per mass decreases with the system pressure of the plasma polymerization reactor. Conversely, the MLCVD process has the unique advantage of yielding nanofilms with very small grain size that have excellent adhesion to the substrate surface, which cannot be obtained by most other modes of plasma polymerization. It has been experienced that most plasma polymer-coated contact lens by conventional RF discharge lose the effect of surface modification when the coated contact lenses are autoclaved, and after two autoclavings, no effect of surface modification was found, indicating the coating could not survive the autoclaving. In contrast to this situation, MLCVD-coated nanofilm of methane showed no influence of surface characteristics after 10 times of repeated autoclaving. These observations are along the line of a tighter network with stronger adhesion to the substrate.

10

Applications of Magneto-Luminous Chemical Vapor Deposition

10.1 Implantation of Imperturbable Surface State on Substrate

10.1.1 Surface Dynamic Change

The topic of surface dynamic change covers a very wide range of case-by-case situations. In this section, only general aspects relevant to magneto-luminous chemical vapor deposition (MLCVD)–prepared surfaces are briefly presented. An extensive review of this topic was presented in reference [1] ("Part III: Fundamentals of Surface and Interface"), which should be able to answer any questions that might arise from reading this abbreviated explanation.

Surface dynamics deals with the change of surface characteristics with time, which is often referred to as *surface reconstruction*. The observation of the surface reconstruction depends on the reference state from which the change occurs, and if one cannot define the reference state, the surface reconstruction cannot be dealt with quantitatively or in a generic sense. Figure 10.1 depicts advancing and receding contact angles of a sessile droplet of water placed on a dry film of ethylene/vinyl alcohol copolymer. The dotted line at Cos θ = 0.7 (contact angle = 45°) is the demarcation line that separates the hydrophilic surface above the line and the amphoteric (partly hydrophobic/hydrophilic) surface, which cannot be considered as hydrophilic or hydrophobic, below the line. The demarcation line between the amphoteric surface and hydrophobic surfaces is Cos θ = 0 (contact angle = 90°) [2,3]. The figure shows that the dry surface of ethylene/vinyl alcohol copolymer film is not hydrophilic, but when contacting liquid water (under the droplet), the surface characteristics progressively shift to the hydrophilic side with contact time with liquid water; after sufficient time, the surface becomes "hydrophilic."

What causes the change from a nonhydrophilic to a hydrophilic surface can be seen by the change in concentration of the O atom (representing –OH group) detected by the angular dependence of the x-ray photoelectron spectroscopy (XPS) atomic ratio of O/C as a function of immersion time of a dry film in liquid water, as shown in Figure 10.2. The data points on the dotted line are values at the top surface of the film, and data points at a higher

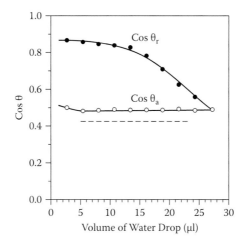

FIGURE 10.1
Advancing and receding sessile droplet contact angle of water on a dry ethylene/vinyl alcohol film.

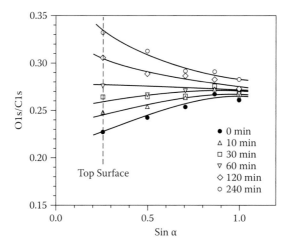

FIGURE 10.2
Change of the concentration of O atom near the top surface of an ethylene-vinyl alcohol copolymer film as a function of water immersion time; XPS angular dependence.

value of Sin α show O/C values in a deeper part of the film. The figure tells us that the concentration of the OH group at the top surface of the dry film is less than that at the inner part of the film, but it changes with immersion time of the film in liquid water and becomes higher than inner parts after 60 minutes of immersion. The water-immersed samples were freeze dried before XPS analysis in order to eliminate the surface dynamic change that could occur due to the change of contacting medium from liquid water to vacuum. The term *surface dynamics* refers to this kind of surface property change caused by the change of contacting medium from air to liquid water in the case dealing with contact angle and XPS angular dependence.

The surface dynamic change is based on the difference from the reference state (starting point); however, the reference state depends on the history of a sample, and the change cannot be reproduced without precise knowledge of the history of a sample [1]. In many cases dealing with surface reconstruction, the reference state is the original state of surface that was created by a film preparation process. In the case of hydrophobic recovery, the reference state is the surface that is just treated to make the surface hydrophilic.

According to the view that a polymeric surface is an ever-changing entity depending on the contacting medium, the restructured surface is not necessarily and is most likely not the final surface (i.e., restructuring of a once-restructured surface or multiple repeated restructuring could occur with highly perturbable polymeric surfaces). The change of surface can be generally explained by the change of surface configuration. Surface configuration refers to what chemical groups actually present at the surface, in contrast to molecular configuration, which refers to what functional groups are in molecules. This distinction can be seen in an example: poly(vinyl alcohol) has an –OH group in a repeating unit according to the molecular configuration, but this does not mean that the surface of the poly(vinyl alcohol) film has all –OH groups or the conformational statistical average at the surface. In reality, a dry surface of a poly(vinyl alcohol) film has less (–OH) groups facing the air phase than one might anticipate based on the molecular configuration; hence, the surface is not hydrophilic.

The general trends with respect to the change of surface configuration caused by the change of contacting medium are reviewed based on the state in which the sample is kept after the change of contact medium. Explaining the surface dynamic changes, the terms *migrate* and *move* are used; however, these motions are within the context of surface configuration. The term *migration* refers to the directional movement of moieties perpendicular to the surface of the film in the context of a surface configuration change, which is achieved by conformational change of macromolecules and should be distinguished from long-range migration such as that by the diffusion transport.

10.1.2 Factors Involved in an Interface

The surface dynamic changes, irrespective of starting points, occur in the direction to minimize the interfacial tension between polymer and the contacting medium. The rate of overall change in the surface-configuration, R_{SCC}, can be viewed as the product of two major parameters; i.e., the polymer chain mobility and the driving force;

$$R_{SCC} = F \text{ (polymer chain mobility)} \times G \text{ (interfacial tension)} \quad (10.1)$$

This is an analogous situation to the diffusive transport of small molecules in a polymer matrix. The flux J can be given by $J = D\ dc/dx$, where D is the diffusion constant (related to the mobility of polymer segments) and dc/dx (concentration gradient) is the driving force;

Diffusion flux = F (polymer chain mobility) × G (concentration gradient) (10.2)

Consequently, any factor that changes polymer chain mobility and any factor that changes interfacial tension cause the change in the rate of surface configuration change.

The following viewpoints are taken into consideration throughout the discussion:

1. A surface can be recognized only as an interface. In this strict sense, only interfacial properties of a solid surface can be considered, and generic surface properties of a solid based on atomic or molecular structure, particularly of a polymer, are misleading.

2. Interfacial properties cannot be described without identifying the contacting medium. Interfacial properties of a polymer solid are dependent on the conditions under which the surface is equilibrated. The surface configuration of a polymer is a function of the (polymer–contacting phase) interface. In this context, the conventional sense of surface property (interface with air) is dependent on the history of the surface and the humidity of air.

3. The surface dynamic change occurs when the interfacial equilibrium is broken and is driven by the interfacial tension in the new environment to establish new equilibrium.

10.1.3 Molecular Configuration versus Surface Configuration

It is generally recognized that the surface of a material is different from the bulk of the same material. However, there are few parameters that could be used to describe differences between bulk and surface properties of a material. With many surface phenomena, such as adsorption of proteins on polymer surfaces, a polymer surface is often treated as if the surface is a rigid and imperturbable plane on which proteins are adsorbed. The configuration of a protein that is adsorbed is one of the most important issues in dealing with the biocompatibility of polymers. However, little attention has been paid to the responsive movement of polymer segments to accommodate a protein in a specific configuration, such as a hydrophilic or hydrophobic moiety facing the surface. The responsive movement of polymer segments can be treated as perturbation of the surface by the contacting phase, which contains a certain kind of protein in the case of protein adsorption. Thus, the adsorption of a protein molecule on a surface of polymer involves both conformational changes of protein and of the polymer that constitutes the surface.

The perturbability of a polymer surface is a function of chemical moieties, which exist on the polymer chain, and the mobility of the polymer chain. The importance of chemical moieties that constitute a polymer is quite obvious, but the role of the mobility of the polymer chain to allow a certain surface configuration is not well understood and is often ignored. There is no argument that poly(vinyl alcohol) (PVA) has many OH groups on a polymer chain, but it cannot be intuitively assumed that all or the majority of OH groups are located on the top surface facing outward. As a matter of fact, the majority of OH groups are not on the top surface of a dry PVA film, as shown in Section 10.1.2.

In recent years, it was gradually recognized that the surfaces of polymers are highly perturbable and that the actual arrangement of chemical moieties of polymer at the surface changes when the surface is brought into contact with a new surrounding medium, for instance, when a polymer surface is immersed in water [4–17]. The change of surface characteristics has been a focal point of the general phenomena recognized by terms such as *surface dynamics*, *surface reconstruction*, *hydrophobic recovery*, and so forth. These terms are reflections of the change of chemical and morphological properties of a polymer surface due to the change of the contacting medium.

Langmuir, in 1938, made the following observations, which pointed out the important aspects of the surface [18]:

1. The wettability of a surface involves only short-range forces and depends primarily on the nature and arrangement of the atoms that form the actual surface and not on the arrangement of the underlying molecular layer.
2. Under special circumstances, a layer at the surface may undergo an almost instantaneous reversal of orientation.
3. Anchoring the reactive groups by forming complexes can prevent such a change.

Adopting the term *surface configuration*, those observations by Langmuir could be rephrased so that the general surface properties could be addressed by the same term:

1. The surface properties of a polymer are determined by the surface configuration rather than the configuration of a macromolecule. It is necessary to distinguish the configuration of a molecule and the *surface configuration* of the same molecule, which is a specific arrangement of atoms or ligands at the surface under a specific condition.

2. Surface configuration changes respond to the conditions under which a surface exists. This is the basic phenomenon that is observed under the topics of *surface dynamics*. The surface configuration is a function of the contacting phase.

3. Surface configuration can be fixed by chemical reactions. This is the foundation for the creation of an imperturbable surface.

Probably the most dramatic demonstration of the importance of *surface configuration* and its distinction from the molecular configuration is the surface characteristics of a hydrogel of gelatin, which contain more than 90% water [16]. It is a general expectation that the surface of such a hydrogel of water-soluble polymer will be highly hydrophilic and easily wettable by water. Contrary to this expectation, the surface of a gelatin hydrogel (gel–air interface) is very hydrophobic, as depicted by the contact angle of water in Figure 10.3. This figure shows that the advancing contact angle as well as receding contact angle of a sessile water droplet is dependent on the humidity of ambient air; that is, the lower the humidity, the higher is the sessile droplet contact angle of water. The high advancing sessile droplet contact angles indicate that the gel–air interface is hydrophobic, but the receding contact angles approach the more hydrophilic side. This difference is caused by the change of surface configuration of gelatin at the interface of the liquid water–gelatin gel surface.

The reason the receding contact angle moves toward the hydrophilic domain becomes clearer when we plot the *contact area* of the sessile droplet against the sessile droplet volume, as depicted in Figure 10.4. During the receding contact angle measurements, the contact area of the water droplet

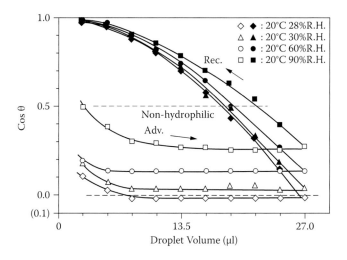

FIGURE 10.3
Effect of droplet volume on advancing and receding contact angles of water on a gelatin gel.

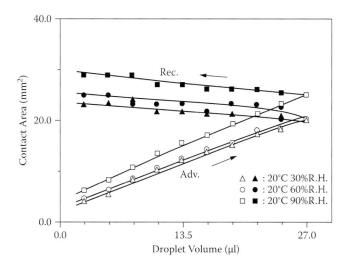

FIGURE 10.4
Effect of droplet volume on the contact area of a water sessile droplet on a gelatin gel.

hardly changes; the reduction of water droplet volume merely flattens the water droplet, causing the contact angle to decline. The strong water droplet holding force by the gelatin–water interface is created by the change of surface configuration of gelatin molecules at the water–gelatin gel interface, which is responsible for these phenomena. Hydrophilic moieties buried in the water-swollen gel phase (95% water) turned around toward the 100% liquid water phase of the sessile droplet. Because the shape of the water droplet changes on the receding contact angle measurement, the value of the receding contact angle is not an accurate measure of the contact angle. But, change of an advancing contact angle to a receding contact angle is a measure of the extent of surface configuration change occurring due to the change of the contacting phase.

In contrast to gelatin gel, a hydrogel of agar-agar that has approximately the same water content shows very hydrophilic surface characteristics as depicted by the similar plots of advancing and receding contact angles shown in Figure 10.5. The advancing and receding sessile droplet contact angles are in the hydrophilic domain, which indicates that the surface configuration at the gel–air interface and the gel–water interface contain enough hydrophilic moieties to make the surface hydrophilic. This means that nearly equal amounts of hydrophilic moieties are present at the interface of gel–air and of gel–water. It further implies that the presence of hydrophilic moieties is not influenced by the nature of the contacting medium (air or water). In the case of gelatin hydrogel, the presence of hydrophilic moieties is greatly influenced by the nature of the contacting medium. In hydrogels that contain more than 90% of solvent (water), the short-range mobilities of macromolecules are high, and the difference between gelatin and agar-agar hydrogels

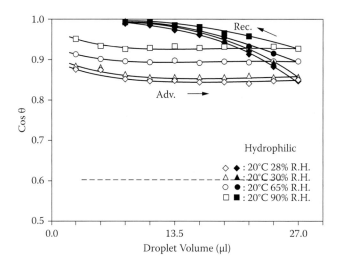

FIGURE 10.5

Effect of droplet volume on advancing and receding contact angles of water on an agar-agar gel.

cannot be explained by the segmental mobility. The difference between the behavior of gelatin hydrogel and agar-agar hydrogen can be explained by the characteristic difference in the molecular configuration of macromolecules.

Why two water-soluble polymers (gelatin and agar-agar) in the hydrogel with the same water content show such a dramatic difference could be explained by the difference of configuration of macromolecules which allows or hinders the surface configuration change. The molecule of gelatin is the denatured collagen (single-strand protein) that takes random conformation in solution and in the hydrogel. The molecule of agar-agar is a polysaccharide, which is shown in Figure 10.6. The molecule of agar-agar also takes the random conformation in the hydrogel; however, the number of surface configurations that can be taken is very small because of the molecular configuration of the macromolecule. First, the repeating unit is planar in which the relative positions of atoms are fixed. Second, the hydrophilic moieties (OH groups) are located on both sides of the plane. The rotation along the axis of chain yields no significant difference in the number of OH groups facing any particular direction. Namely, if an OH group is moved from the surface by the rotation, another OH group on the other side of the plane comes up to the surface.

The differences in the surface characteristics of gelatin gel and agar-agar gel can be explained by the orientation of hydrophilic moieties at the surface as shown in Figure 10.7. In the case of gelatin gel, the hydrophilic moieties prefer to stay in the aqueous phase, leaving the hydrophobic sections of a macromolecule on the top surface in contact with air. The orientation changes immediately when the gel is placed in water or is taken out of water because of the high mobility of segments in the low concentration (highly hydrated

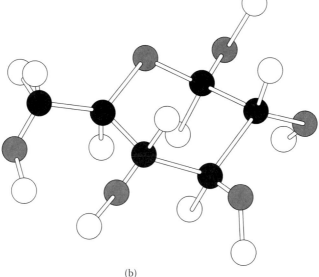

(a)

(b)

FIGURE 10.6
The molecular configuration of the repeating unit of agar-agar.

gel). Consequently, if the contact angle is measured at the gel–water interface, rather than at the gel–air interface, the sessile bubble contact angle is
low, indicating that the same gel shows very hydrophilic characteristics.

In the case of agar-agar gel, in which the double-sided model represents
the hydrophilic moiety, the number of the hydrophilic moieties at the air–gel
interface and at the water–gel interface is nearly the same. Consequently, the
sessile droplet contact angle of water is low and nearly identical to the sessile
bubble contact angle of water.

MLCVD processing is highly useful in creating a fixed surface configuration. A *surface configuration* could be fixed to yield an imperturbable or less
perturbable surface. Whether or not the surface of a polymer is hydrophilic

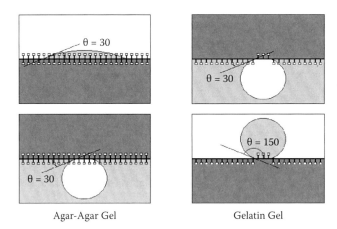

Agar-Agar Gel   Gelatin Gel

FIGURE 10.7
The effects of the orientation of hydrophilic moieties on the sessile droplet contact angle and sessile bubble contact angle.

or hydrophobic is not determined by whether or not hydrophilic or hydrophobic moieties exist in a polymer molecule but is determined by what kind of moieties actually occupy the top surface of the polymer at the polymer–air interface. Namely, the surface properties are controlled by the *surface configuration* but not by the configuration of polymer molecules. The configuration of PVA can be represented by $-[CH_2-CH(OH)]_n-$, which indicates that the polymer is highly hydrophilic, and PVA is water soluble. However, the surface characteristics of a film of PVA depend entirely on how many OH groups exist on the top surface or are buried in the bulk phase of the polymer, particularly in the crystalline phase. Instead of forming complexes or relying on the presence of the crystalline phase, a very tight covalently bonded network of amorphous carbon film formed by MLCVD is used in the approach described in this book.

10.1.4 Implantation of Surface State of Magneto-Luminous Chemical Vapor Deposition (MLCVD) Nanofilm onto Material Surface

Three unique and important features of *Type-A* MLCVD nanofilm make MLCVD coating an ideal tool in the preparation of biomaterials. They are an imperturbable surface, a nanoscale molecular sieve, and a new surface state of material. It should be reiterated that these three features of MLCVD films are limited to *Type-B* plasma polymers as described in Chapter 8; *Type-A* plasma polymers should be excluded for biomaterials based on the concept of imperturbable surface. Particularly important is that the MLCVD nanofilm becomes the new surface state of the substrate material; that is, it is not just a coating placed on the surface. The first and the second features describe the nature of the new surface state.

As far as the biocompatibility of a man-made material is concerned, the "imperturbable" aspect of the surface describes the way that luminous chemical vapor deposition (LCVD) coating could be best utilized in biomaterials. The interaction, more precisely the absence of it, of a man-made material (not existing in the natural state of a biological system) with a biological system is generally expressed as *biocompatibility*. The materials that have biocompatibility are referred to as *biomaterials*. However, the true sense of biocompatibility cannot be described without specifying the details of the biological system and how the biomaterial is placed in the biological system.

The biocompatibility that could be attained by MLCVD coatings could be illustrated by the terms used to describe *blood compatibility*, which is a vague term that indicates how well a material could be allowed to contact with blood without causing the coagulation of blood. The interaction of a material surface with blood could be described by terminology that describes the consequence of blood–material contact, such as a thrombogenic surface, which causes the coagulation of blood, and a nonthrombogenic surface, which does not cause the coagulation of blood. The nonthrombogenic surface could be further divided into the antithrombogenic surface, which positively prevents the coagulation of blood, and athrombogenic surface, which does not cause the coagulation of blood but does not prevent the natural coagulation of blood either. An example of an antithrombogenic surface is the heparinized surface, on which the anticoagulation agent, heparin, is incorporated chemically or physically. The heparinized surface prevents the coagulation of blood very effectively and solves acute coagulation problems associated with the use of biomaterials.

Even though a biologically active surface performs well based on the specific biological reaction, it is a highly perturbable surface tailored for a specific reaction that could, in principle, cause other biological reactions. For instance, a heparinized surface seems to increase hemolysis (breakdown of red blood cells). It should be considered that when the biologically active agents wear out, the surface of the treated material returns to the untreated surface, which needed the addition of an anticoagulation agent in the first place. The sustainability or durability of an antithrombogenic surface is a key issue that is often ignored in biomimetic approaches.

In contrast to the antithrombogenic surface, an athrombogenic surface does not participate in any biological interaction; that is, it does not cause blood coagulation or prevent coagulation, and it stays neutral in the biological environment. Such an ideal surface cannot be created by the surface modification of existing materials by conventional means of chemical reaction and physical treatment, because those means require the surface to be highly perturbable; consequently, the product would be highly perturbable. It is important to recognize that the adsorption of a protein on a material surface, which is a relatively simple interaction of a biological component with the surface, involves the perturbation of the surface, such as the change of the surface configuration.

The surface is a crucially important factor of a biomaterial, and without appropriate biocompatibility, the biomaterial could not function. On the other hand, the bulk properties of materials are equally important in the use of biomaterials. An opaque material cannot be used in vision correction, and soft flexible materials cannot be used in bone reinforcement. The probability of finding a material that fulfills all requirements in physical and chemical bulk properties needed in a biomaterial application and the surface properties to be just right for a specific application is virtually zero. From this point of view, all biomaterials should be surface treated to cope with the biocompatibility. If the surface treatment alters the bulk properties, however, it defeats the purpose. In this sense, a tunable MLCVD nanofilm coating that causes the minimum effect on the bulk material is the best tool currently available in the domain of biomaterials.

The imperturbable surface is related to the number of surface configurations that can be taken at the surface and its change when the surface contacts a biological system. The number of surface configurations is governed by the molecular configuration of a polymer molecule. This point can be visualized by comparing the molecular configuration of poly(oxy methylene) (POM) and that of poly(oxy ethylene) (POE). Top views, side views, bottom views, and end views of a stretched short segment of POM and POE are compared in Figure 10.8. Although such a stretched conformation is unlikely to be found in a POM film, the figure illustrates the meaning and its importance of surface configuration. Figure 10.8 compares only a specific conformation, the stretched form, of the polymer segments, but it indicates the significant

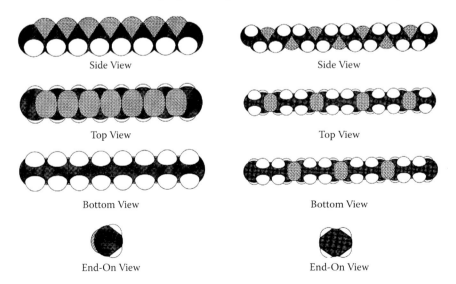

FIGURE 10.8
Comparison of stretched short segment of POM (left) and POE (right) in top, side, bottom, and end views.

difference in the number of surface configurations that could be taken by these two polymers. Because of the symmetry of configuration along the chain axis of POE, the change of surface configuration due to the change of contacting medium is anticipated to be small. This is a similar situation as described for the difference of sessile droplet contact angle which indicates the interfacial tension at the solid–air interface and the sessile bubble contact angle that indicate the interfacial tension at a solid–liquid interface for gelatin and agar-agar hydrogels. The surface of POE is virtually imperturbable by virtue of its molecular configuration; that is, the surface configuration remains nearly the same regardless of short-range conformational and rotational changes. It is known today that the POE surface is one of the most biocompatible surfaces.

According to the molecular configuration of repeating units; $[-(CH_2-O)_n-]$ for POM and $[-(CH_2CH_2-O)_n-]$ for POE, O/C ratio of POM is twice of that of POE, which might suggest POM is more hydrophilic than POE. Contrary to the speculation, the surface of POM is hydrophobic, which means that oxygen atoms are not contributing to the surface configuration (mostly not exposed to the top surface), and the surface can be represented by the bottom view shown in Figure 10.8. It is important to recognize that the molecular configuration (i.e., $-(CH_2-O)_n-$) is fixed, and the three different surface configurations under consideration are the result of different orientations of the fixed conformation of POM shown in the figure. The figures simply demonstrate that the surface configuration can be changed by a simple mode of rotational and conformational change of a polymer segment. The actual surface configurations that can be taken by POM molecules are much more complex, because the surface configuration can be altered by changing the conformation of macromolecules. (The conformation of the segment is fixed in the figure.) Figure 10.8 clearly shows that POM could take many surface configurations, which indicate that the surface of POM could be highly perturbable.

A striking difference between POM and POE is in the possibility of taking various surface configurations. As can be easily visualized from the figure, the rotation along the axis of the chain of POE yields very little change in the surface configuration, whereas the same action on POM yields numerous surface configurations. Thus the POE surface is virtually imperturbable, and POE is a hydrophilic water-soluble polymer, while POM is a hydrophobic water-insoluble polymer. The surface of POE is virtually imperturbable by virtue of its molecular configuration; that is, the surface configuration remains nearly the same regardless of short-range conformational and rotational changes. The POE surface is one of the most biocompatible surfaces.

In contrast to the case of POE, the surface of MLCVD-formed amorphous carbon nanofilm (plasma polymer) is imperturbable by virtue of the lack of molecular mobility at the surface (i.e., amorphous carbon network). It should be cautioned, however, that the imperturbable surface is limited to Type-A plasma polymers and does not extend to any plasma polymers with discernible functional groups. The details of biocompatibility are beyond the scope of this book. Only some data that show good biocompatibility of the surfaces of MLCVD coatings are shown in this chapter.

It should be emphasized, however, that these examples clearly demonstrate the validity of the neutral approach explained by the athrombogenic surface in Section 10.1.4. If we could reduce the overall perturbation to the host biological system, caused by an artificial material, to within the threshold level that can be tolerated by the host biological system, the artificial material can be tolerated by the host biological system—the implant is biocompatible. The minimum perturbation principle [19] is not in line with the biomimicking approach. In fact, it is a completely opposite approach to the biomimicking principle; that is, the absence of any chemically and biologically active functional group is the key factor for the biocompatibility of artificial materials. Such an approach can be effectively achieved with the MLCVD process.

10.2 MLCVD Nanofilm for Biocompatibility

10.2.1 Imperturbable Surface State and Biocompatibility

The imperturbability of a surface is a crucial factor, in the author's view, in the biological compatibility of artificial surfaces. The imperturbability surface could be attained by molecular configuration of a macromolecule, such as the case of poly(oxy ethylene), or an immobile network structure in the top surface, such as the case of the tight network of MLCVD nanofilm.

The change of surface configuration occurs in order to minimize the free energy difference at the interface, which is recognized as the interfacial tension. It is important to recognize that this interfacial change precedes any change that occurs in the subsurface layer or in the bulk of the material. The change of surface characteristics is not necessarily a consequence of long-range morphological or conformational changes of macromolecules. Short-range motions, such as rotation along the axis of a polymer chain and short-range conformational change, could alter surface configuration.

With consideration of surface properties as a function of the contacting phase or the surrounding medium, some serious questions arise: "What is a surface?" "What is an interface?" In recognition of the fact that surface properties depend on the surrounding medium, including air or vacuum, one comes to a conclusion that the surface (exterior of an object) always exists as an interface with a surrounding medium, which could be in gas, liquid, or solid phase. The surrounding medium includes ambient air and vacuum. The interface with ambient air or with vacuum is customarily recognized as the surface, and surface characteristics are expressed by the properties of the interface with air or vacuum.

The surface configuration is a function of the nature of the contacting medium (environment) as well as of temperature. The arrangement of surface atoms under consideration changes by virtue of the change of conformation of macromolecules, but the configuration of macromolecules remains unchanged. The changes of interfacial characteristics in real situations are further complicated by the contributions and their changes of surface impurities and low molecular weight components, which tend to accumulate on the surface.

In a strict sense, there are no surface characteristics that can describe behavior of a material in a biological system. Every surface is an interface with a specific contacting medium, and interfacial characteristics cannot be described without specifying the contacting medium. The phenomena recognized by terms such as *surface dynamics*, *surface reconstruction*, and *hydrophobic recovery* reflect on the changes of the reference state, because surface characteristics are generally measured with interface with air or vacuum. In other words, how a synthetic polymer behaves in contact with blood or biological tissues cannot be predicted from the interfacial characteristics of the materials with air or vacuum alone. The biological system is so much different from a simple contacting phase (e.g., vacuum, air, water, etc.) that interfacial properties become the most important factor in comprehending the biocompatibility of artificial materials.

10.2.2 Encapsulation of Silicone Contact Lens

The three key features of MLCVD coating ideally suited for a biomaterial surface and the important balance between the bulk properties and the surface properties could be illustrated by examples of nanofilms of MLCVD methane plasma polymer on the contact lens made of elastomers of poly(dimethylsiloxane). Some details of processing factors and their influence on the overall properties of the product are described in the following sections.

Silicone rubber, poly(dimethylsiloxane), has great advantages as a contact lens material because of its very high oxygen permeability, softness, and good mechanical strength and durability. However, practical utilization is hampered by inherent surface characteristics of elastomers (i.e., high tackiness and highly hydrophobic surface properties). The characteristically high permeability of the material to various molecules also accounts for several problems encountered with silicone contact lenses. When a silicone rubber contact lens is placed on the cornea, many lipid-soluble materials in the tear film adhere and eventually penetrate into the bulk phase of the lens, resulting in degradation in optical clarity, which cannot be eliminated with routine lens cleaning. The characteristically high permeability due to the high free volume means, on the other hand, that silicone rubber is one of the poorest barrier materials, not only for gases and water vapor but also for lipids and lipid-soluble materials.

Liquid water on one side of a silicone contact lens permeates through the lens by the solution-diffusion mechanism and evaporates on the other side according to the permeability of water, whereas the solubility of water in a silicone polymer is low [20]. The high water vapor permeability was speculated as one of the reasons causing the "suction cup effect" that makes the lens stationary on one spot while tenaciously sticking to the cornea; this may damage the corneal epithelium and result in other complications. However, the high permeability, per se, cannot be the reason for the suction cup effect, if the exterior surface is covered by the tear film; that is, there is no driving force for water permeation across the interface, and a layer of tear fluid prevents strong adhesion (suction cup effect).

The hydrophobicity of the surface prevents the wetting by tear and tends to expose the dry surface of a contact lens. Therefore, rapid dehydration of the corneal tissues could occur, which could cause damage to the corneal epithelium. However, this explanation seems to be oversimplified, in light of the adsorption of protein, which makes a hydrophobic surface wettable by tear fluid. Moreover, the highly hydrophobic surface characteristic of silicone rubber tends to encourage the deposition of protein and mucus of the tear on the surface of the lens. Lipids and lipid-soluble materials follow the same track of the diffusive permeation and eventually penetrate into the bulk phase of the contact lens. Because of these undesirable factors, silicone contact lenses of various chemical compositions and with surface treatments have not only been unsuccessful, but they have actually been disastrous because of the interfacial characteristics of silicone contact lens on the cornea, which cannot be changed by efforts to improve surface characteristics of silicone polymers. This indicates that more profound surface modifications rather than mere surface treatments are necessary to cope with the problems of capitalizing the great, advantageous, bulk properties of silicone polymers.

It should be considered that the modification of the molecular configurations of polymer segments can change the bulk properties of polymers but is rather ineffective in changing the surface property of a polymer, particularly in imparting contradicting properties such as a hydrophilic surface on a hydrophobic polymer or a hydrophobic surface on a hydrophilic polymer, as schematically depicted in Figure 10.9. This strongly indicates the need and merit of application of a nanofilm on the surface, or more specifically the implantation of surface state with characteristics different from the bulk phase of the polymer.

The major problems associated with silicone rubber used for contact lenses stem from the surface properties. The surface is hydrophobic and hinders the spontaneous spreading of tear on the surface. The surface of silicone rubber has tackiness that is characteristic of elastomers. Those two factors together are probably the main reason why silicone lenses tend to stick on the cornea. MLCVD of CH_4 was applied on the surface of a silicone contact lens and flat

Influence of Chemical Structure of
Polymer on Surface Properties

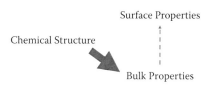

Surface Properties

Chemical Structure

Bulk Properties

Molecular Configuration Controls the Bulk Properties
but not Surface Property

FIGURE 10.9
Factors that control bulk properties and surface properties.

sheet of the same material to investigate the effect of coating on the stickiness of the silicone surface (change of surface state) and the imbibitions of lipids and lipid-soluble material by silicone polymer (barrier characteristics of the coating) by means of the absorption of oil-soluble dye as a function of the operational parameter of CH_4 MLCVD nanofilm placed on the surface of silicone contact lens [21,22].

10.2.2.1 Dye Penetration Test

Due to the high free volume, silicone rubber absorbs oil-soluble organic dyes easily, and the oil–soluble dye absorption test is an effective indicator in evaluating the uniformity and the barrier characteristics of the coating. When a droplet of oil-soluble Sudan Red in butanol is placed on the interior surface of a silicone contact lens (a lens as a cup) for several minutes and rinsed by butanol, the lens shows a bull's-eye of red stain. This means that the oil-soluble dye penetrated the bulk of the silicone contact lens in the short contact time. When the stained lens is kept overnight, the brilliant red bull's-eye is transformed to a lightly stained contact lens, because the dye penetrated from the droplet of dye solution placed on the surface which migrated to the entire contact lens. These observations show two important characteristics of the silicone contact lens: high absorption characteristic to oil-soluble material and high diffusivity of the absorbed dye molecules within the bulk phase of the contact lens.

If the coating applied is a good barrier to the dye, no stain is observed. If pinholes exist, spot staining occurs, and if the coating is permeable to the dye molecule, the diffused staining occurs, of which intensity is inversely proportional to the barrier characteristic of the coating. Moreover, if plasma film can prevent the Sudan Red dye molecules (MW 380.48) from penetrating the highly permeable silicone rubber contact lens (no stain), it can be safely considered that the coating is an effective barrier to the diffusion of larger lipids and lipid-soluble molecules in tears. Thus, the simple dye test serves as a crucially important test that provides information pertinent to the integrity and barrier characteristics of LCVD nanofilm placed on a silicone contact lens. As described in Chapter 9, it was recently found that the operational parameter of MLCVD does not follow precisely with the W/FM parameter; however, the general trends of material formation still follow the same trends based on the W/FM parameter; hence, the data expressed as functions of W/FM are presented here without recalculating the parameter.

Table 10.1 summarizes the results of dye tests of plasma polymerization coatings of methane and a mixture (methane + nitrogen) prepared under various conditions. In this table, the plus sign (+) indicates that the sample passed the test, and the minus sign (−) means the sample failed the test. MLCVD polymerized methane coating thickness, as read from the thickness monitor, was 12 nm for all coatings shown in the table. It shows that for a pure

TABLE 10.1

Dye Test for Contact Lenses Clamped on Rotating Disk and Coated with Methane Plasma Polymer or Methane Plasma Copolymer at Power of 75 W

Flow Rate of CH$_4$ (sccm)	W/FM (GJ/kg)	Gas Mixed with CH$_4$ to Initiate Plasma	Treatment after CH$_4$ or Mixed CH$_4$ Plasma	1	2	3	4	5	6
0.29	21.8			+	−		−		
0.29	21.8			+	−		−		
			(half in dye)						
1.64	3.8			+	+	+	−	+	+ −
3.28	1.9			+	+	+	+	+	+ −
2.46	2.6			+	+	+	+	+	+
2.46	2.6		Immersed in 70°C water for 10 min	+	+	+	+	+	+
2.46	2.6		O$_2$ plasma, $F = 10$ sccm, $W = 12.5$ W, 2 min	+	+	+	+	+	+
2.46	2.6		Wet air passing	+	+	+	+	+	+
2.46	2.6		Wet air plasma $W = 50$ W, $F = 10$ sccm, 2 min, followed by immersion in 25°C water (using lens that can pass the 5-min dye test)	+	+		+		
2.46	0.8	N$_2$, F = 3 sccm		+	+		−	+	+ −
2.46	0.8	N$_2$, F = 3 sccm	Immersion in 70°C water for 10 min	+	−	−	−		+ −
2.46	1.5	N$_2$, F = 1 sccm	Immersion in 25°C water	+	+		−		
2.46	1.5	Air, F = 1 sccm	Immersion in 25°C water	+	+		+		

Notes: The thickness reading was 12 nm. All the coated contact lenses were dipped in dye for 5 min. Dye test: 1, no mechanical work; 2, rolling once; 3, rolling three times; 4, rolling 10 times; 5, slight folding; 6, bending, +, no red stains appeared; −, red cloud appeared; + −, few red stains appeared.

methane coating at $W/FM = 2.6$ GJ/kg, the various kinds of mechanical resistance of the coating film are better than those prepared in other conditions (i.e., $W/FM = 21.8$ GJ/kg). This implies that at $W/FM = 2.6$ GJ/kg, the adhesion of the methane plasma polymer to silicone lens, and the cross-linked network of methane plasma polymer, are optimal. The coating prepared at the higher W/FM seems to be too brittle and failed under mechanical stress.

10.2.2.2 Effect of Coating Thickness

Tables 10.2 and 10.3 depict the effect of coating thickness on the dye penetration tests for coating prepared at 2.6 GJ/kg and 21.8 GJ/kg, respectively. Holding a coated contact lens folded between the thumb and the index finger tightly and rolling the folded contact lens back and forth repeatedly applied flexing of the coatings. The results of the dye tests are related to the internal stress of the coating film, which might cause cracking of the coating, and the adhesive strength between the plasma coating and the substrate material. Adhesive strength provides the positive effect to get a tenacious coating on the substrate, whereas the internal stress causes cracking of the film. The rolling of a coated contact lens for 10 times was used to test the mechanical

TABLE 10.2

Results of Dye Test of Contact Lenses Coated with Methane Plasma Polymer at $W/FM = 2.6$ GJ/kg

Reading Thickness (nm)	Coating Thickness (nm)	5-Min Dye Test	
		No Mechanical Work	Flexing 10 Times
10	2.5	+	−
12	3.0	+	+
20	5.0	+	+
25	6.2	+	+
50	12.5	+	+
75	18.8	+	+
100	25.0	+	+
125	31.2	+	+
150	37.5	+	+
175	43.8	+	+
200	50.0	+	−
250	62.5	+	−
300	75.0	+	−
350	87.5	+	−
400	100	+	−
440	100	−	−

Notes: All the contact lenses were clamped on the edges of empty holes of a rotating disk to be coated. All the coated lenses were dipped in dye for 5 min. Power = 75 W; flow rate = 2.46 sccm; +, no stain appeared; −, red cloud appeared.

TABLE 10.3

Results of Dye Test of Contact Lenses Clamped on the Rotating Disk
and Coated with Methane Plasma Polymer at W/FM = 21.8 GJ/kg[a]

Reading Thickness (nm)	Coating Thickness (nm)	5-Min Dye Test	
		No Mechanical Work	Flexing 10 times
10	3.5	−	−
12	4.2	+	−
20	6.9	+	+ −
30	10.4	+	+ −
50	17.3	+	+ −
70	24.2	+	−
150	51.9	+	−
200	69.2	+	−
250	86.5	+	−
300	103.8	+	−
350	112.1	+	−
400	138.4	+	−

Notes: Power = 75 W; flow rate = 0.29 sccm; +, no stain appeared; −, red cloud appeared;
   + −, few red stains appeared.
[a] All the coated contact lenses were dipped in dye for 5 min.

breakdown of the coating. Comparing Tables 10.2 and 10.3, it is clear that at
a W/FM of 2.6 GJ/kg, a well-covered pin-hole-free uniform coating can be
obtained from 2.5 to 100 nm, and mechanical-resistant film can be obtained
from 3 to 44 nm. On the other hand, at a W/FM of 21.8 GJ/kg, a well-covered
flawless film is obtained from 4.2 to at least 140 nm, but a mechanically resistant coating film was never achieved.

When coating thickness is less than the minimum threshold values (2.5 nm
at low W/FM, 4.2 nm at high W/FM), it is impossible to get a well-covered
uniform coating film; therefore, the dye penetrates into the contact lens and
stains appear. As the coating thickness gradually increases, the uniform film
becomes well developed. Nevertheless, the internal stress simultaneously
increases with the coating thickness until it reaches the threshold value and
cracking of the film occurs. Therefore, the barrier characteristic is lost, and the
red stains show up.

When an external flexing force is applied on the coated substrate, the film
of W/FM 21.8 GJ/kg cannot maintain the integrity of the coating. The coating
of W/FM 2.6 GJ/kg has a 35 nm durable range, which means that the coating film of low W/FM has more ability to resist mechanical work than that
of high W/FM. Compared with the low-energy film, the high-energy film is
accompanied by higher internal stress. The cohesive forces of all the coating
films of various thickness at W/FM = 21.8 GJ/kg cannot withstand the total
force of internal stress and external force, and thus, cracking results. The
cohesive force of the film of W/FM = 2.6 GJ/kg can stand the total force until

the coating thickness reaches 44 nm. At that point, the cohesive force is over-come by the combination of internal stress and external force. These results clearly demonstrate that so far as the quality of MLCVD coating as a func-tion of discharge energy is concerned, the principle of "the more the better" definitely does not apply, and the identification of the domain of MLCVD is crucially important.

10.2.2.3 Effect of Power Input Level

The effects of the composite parameter *W/FM* on the friction coefficient of the contact lens coated by methane plasma polymer at a fixed coating thick-ness 31.2 nm are shown in Figure 10.10. The friction coefficient, which is the tangential force divided by the normal force, was calculated from the sliding angle of a contact lens placed on a glass plate that was coated with plasma polymer of tetrafluoroethylene (TFE). The friction coefficient of the MLCVD methane plasma polymerized on the contact lens is independent of *W/FM* in the range of 2.6 GJ/kg to 29.1 GJ/kg.

Figure 10.11 depicts the change of frictional coefficient as a function of coating thickness at *W/FM* = 2.6 GJ/kg, which indicates that the applica-tion of the MLCVD coating effectively reduces the frictional coefficient, and above thickness of roughly 15 nm, the frictional coefficient is independent of the coating thickness. The uncoated silicone rubber contact lenses are very sticky, with a falling angle of greater than 90°. But after the silicone rubber contact lens is coated with 2.5 nm of methane MLCVD CH_4 polymer, the fall-ing angle decreases to 78°. In other words, the friction coefficient is 4.7. When coating thickness is between 2.5 and 25 nm, the friction coefficient decreases very fast, as does the tackiness. Over 25 nm, the friction coefficient is inde-pendent of coating thickness and becomes stable around 0.4.

Many factors influence tackiness, such as humidity, surface properties, elas-ticity, and the shape of the tested material. The most plausible explanation

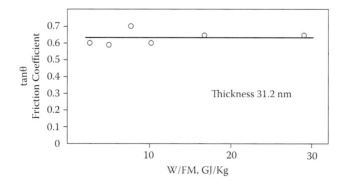

FIGURE 10.10
Effect of discharge energy input on the frictional coefficient of coating.

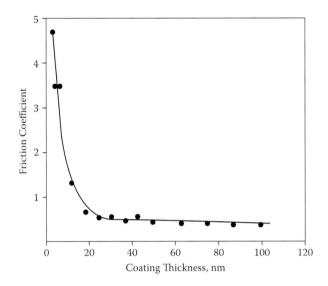

FIGURE 10.11

Effect of coating thickness on the frictional coefficient of a silicone contact lens coated with MLCVD CH_4 plasma polymer.

for the reduction of friction coefficient by MLCVD coating of methane on the contact lens is that the coating changes the surface state of the silicone rubber. The thickness dependence agrees with that for the necessary thickness to change the surface state. Although it cannot be expressed numerically, a temporary coating dramatically changes the feel of the surface from "tacky" to "slippery."

The dependence of the falling angles of contact lenses on the coating thickness at a *W/FM* of 2.6 GJ/kg and the results of the dye tests in Table 10.2 showing uniformity and mechanical resistance are combined in Figure 10.12. This shows that the uniform and pin-hole-free film is in the range of coating thickness from 2.5 nm to 100 nm, while the mechanical resistant film is in the range of 3 nm to 43.8 nm thick. This indicates that a MLCVD coating of methane, of which thickness is roughly 3 nm to 40 nm prepared under the low energy input condition (2.6 GJ/kg normalized energy input), satisfies the requirements for obtaining uniform, flawless, highly mechanically durable film with an appropriate friction coefficient (without tackiness). It should be emphasized, however, that these conditions are dependent on the nature of the substrate and should not be taken in a generic sense. The important point is that the best results are obtained with a minimum thickness of coating prepared under a mild condition of the MLCVD process; the principles of "the thicker the better" and "the higher energy input the better" do not apply in the surface-state implantation process.

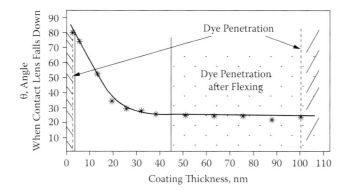

FIGURE 10.12
The range of coating thicknesses that satisfy both frictional coefficient and the dye penetration test at W/FM 2.6 GJ/kg.

At the time of investigation in the mid-1970s, it was not clear why lower power operation yielded much better nanofilms; however, with the knowledge of fundamentally different mechanisms of dielectric breakdown of the gas phase in low pressure under the influence of a magnetic field, found recently and described in this book, it is the uniquely advantageous factor of low-voltage/high-current discharge in the [M] domain of the magneto-luminous gas phase.

10.2.2.4 Overall Effects of MLCVD Coating

Properties of MLCVD CH_4 coated (thickness 5 nm) silicone contact lenses are compared with those for uncoated lenses in Table 10.4. In the coating process designated as Condition C, the energy input at 2.6 GJ/kg was used (F = 2.46 sccm, 75 W), and wet oxygen was introduced into the reactor after the polymerization and evacuation of gas. It was intended to enhance the reaction of dangling bonds with oxygen and water to yield a slightly more hydrophilic surface than the normal exposure to ambient air. The post-deposition oxygen plasma treatment of the coating was found to reduce the contact angle of water significantly, but at the expense of surface dynamical stability; hence, it was not adopted in the sample preparation.

The coating near the low end of coating thickness satisfies both the dye penetration test and the reduction of stickiness of the surface. The CH_4 MLCVD coating at the thickness of 5 nm effectively removed all problematic features of silicone contact lenses with a 12% reduction in oxygen permeability, which is still much higher than that of most other materials, due to the nanoscale molecular sieve aspect of MLCVD film. The true significance of this coating is the change of surface state of the silicone lens to the imperturbable surface

TABLE 10.4

Comparison of Properties of Coated Contact Lenses and Uncoated Contact Lenses in Condition C

	Uncoated	Coated
Wettability (degree; dyne/cm)		
Contact angle of water, θ	81	60
		(26% reduction)
Dispersion component of surface energy	22.54	26.96
		(19.6% increase)
Polar component of surface energy	5.63	12.50
		(122% increase)
Solid surface energy	28.17	39.46
		(40% increase)
Permeability $(cm^3 - STP)$ $(cm)/(cm^2)$ (s) $(cm\ Hg)$		
O_2	557×10^{-10}	508×10^{-10}
		(12% reduction)
N_2	282×10^{-10}	193.3×10^{-10}
		(31% reduction)
CO_2	2442×10^{-10}	2046×10^{-10}
		(16% reduction)
Tackiness		
Falling angle of contact lens	90	75
Friction coefficient (tan θ)	∞	3.73
ESCA		
O_{1s}/C_{1s}	0.46	0.271
S_{i2p}/C_{1s}	0.55	0.195
Dye test	Stain	No stain
Softness	Good	Good
Transparency	Good	Good

state of MLCVD CH_4 nanofilm. Although evaluations were positive, there is no commercially available MLCVD CH_4 encapsulated silicone contact lens; probably due to the previous bad reputation of uncoated silicone contact lenses and lack of awareness of the distinction between surface modification and the encapsulation of a whole contact lens. Above all, however, the current fashionable trend in contact lens manufacture is silicone hydrogel, and that seems to be swept away by the potential of silicone contact lens (without a hydrogel component). All industrial-scale applications of MLCVD CH_4 coatings are with silicone/hydrogel contact lenses, which are described in the next section.

All results shown in Section 10.2.1 are totally in line with the recent understanding of MLCVD in the [M] domain of the gas phase breakdown phase diagram described in Chapter 7, although the data were obtained over 23 years ago, which attests to the robustness and reliability of the MLCVD process.

10.2.3 Encapsulation of Silicone/Hydrogel Contact Lens

10.2.3.1 Advantages and Disadvantages of Silicone/Hydrogel Contact Lenses

Contact lenses were developed after World War II by machining poly(methyl methacrylate) rod in the form of a small button. It was said that the use of poly(methyl methacrylate) (PMMA) as a contact lens was intriguing because small fragments of the canopy of a war plane remained in the cornea of a pilot without causing any foreign body reaction. The use of a hydrogel (swollen hydrophilic polymer) for contact lens material, instead of hard PMMA, literally revolutionized the practical use of contact lenses in the mid-1960s because of the ease of mass production by casting hydrophilic monomers for hydrogel in molds or in spinning to polymerize the shapes of contact lenses, instead of machining each individual lens; the ease of wearing them, because of the softness of the lenses; and the hydrophilic surface of contact lenses, which contributes to the initial wearing comfort. However, wearing them for a longer time causes a hypoxia condition to form at the cornea, which has been attributed to the low oxygen permeability of hydrogel contact lenses. Regardless of whether this assessment is correct or not, the permeability of oxygen through a hydrogel contact lens is low, because oxygen permeates through liquid water in the swollen hydrogel polymer. The diffusivity of oxygen in liquid water, D, is high; the solubility of oxygen in liquid water, S, is low; and the overall permeability constant, P, which is given by $P = D * S$, is considered not high enough for a contact lens.

In order to overcome the shortcoming of hydrogel contact lenses, silicone/hydrogel contact lenses were developed by taking advantage of the exceptionally high (among conventional polymers) oxygen permeability of silicone polymers. Contrary to the oxygen permeability in liquid water, silicone polymers have much higher solubility of oxygen in solid polymer than in liquid water. Although the diffusivity of oxygen in a silicone polymer is smaller than that in liquid water, the permeability of oxygen through silicone rubber is seven to eight times higher than that in liquid water. Gases permeate through a polymer by first being dissolved in the free-volume in a polymer and then diffuse through hopping from one free volume to other free volumes driven by a chemical potential gradient. Silicone polymers (rubbers) have exceptionally high free volume among most commercially available polymers.

A silicone hydrogel material is synthesized by polymerizing a hydrophilic monomer, or monomers, in the presence of a low molecular weight polymer of silicone that has functional groups on each end of its segment or segments, which act as initiators of the polymerization of the hydrophilic monomer. Table 10.5 lists typical commercially available silicone hydrogel contact

TABLE 10.5

Commercial Silicone Hydrogel Contact Lenses

Product Name	FND	Pure Vision	Acuvue Advance	O2 Optix	Phenix	Acuvue (1&2)
Manufacturer	Ciba	B&L	J&J	Ciba	J&J	J&J
USAN	Lotrafilcon A	Balafilcon A	Galyfilcon A	Lotrafilcon B	Senofilcon A	Etafilcon A
Water content (%)	24	36	47	33	38	56
Oxygen permeability (Barrer)	140	99	60	110	103	20
Modulus (MPa)	1.7	1.1	0.5	1.0	n/a	0.35
Polymer	Acrylate	Vinyl	Acrylate	Acrylate	(Meth) acrylate, vinyl	Methacrylate
Surface	Plasma polymer coating	Plasma oxidation	Entangled PVP	Plasma polymer coating	Entangled PVP	None
Hydrophilic components	DMA	NVP, amino acid monomer	DMA, HEMA, PVP	DMA	DMA, HEMA, PVP	HEMA, MA

lenses. As seen in the table, silicone/hydrogel contact lenses require compromising at least two major characteristics: soft hydrophilic surface of hydrogel with low oxygen permeability and high oxygen permeability of silicone rubber with a hydrophobic surface.

At this point, it might be worth examining key factors of contact lens wearing, instead of focusing on the two major factors involved in silicone/hydrogel materials. The following factors in the wearing comfort of a contact lens are important in selecting a suitable contact lens for a wearer:

1. Initial wearing comfort in a specified time period (e.g., of 1 hour): The initial wearing comfort is the most important factor for new users of contact lenses. The soft hydrogel lenses have superior advantage in this factor. The water wettable surface and the softness of lens material are key factors. If the initial wearing comfort is most important, hydrogel contacts should be worn.

2. The secondary wearing comfort after the initial period of wearing: The wearing comfort of hydrogel contact lenses declines with wearing time, and when the wearer unintentionally closes his or her eyes for some period of time (dozing), the vision declines significantly when the wearer wakes up. Hypoxia in the cornea develops in this stage.

3. Unintentional overnight wearing: This seems to happen more frequently than anticipated. Hydrogel contact lenses should not be worn overnight; furthermore, the removed contact lenses must be kept in a cleaning solution overnight. Potential infection is the most serious drawback of wearing contact lenses.

4. Extended wearing of contact lenses: According to U.S. Food and Drug Administration (FDA) regulation, some contact lenses could be worn continuously for 1 week, and some could be worn for 1 month without removal. The infection issue is least with use of longer wearing contact lenses, if the initial insertion of the contact lenses was done appropriately.

One major drawback of silicone/hydrogel contact lens is the following: If the silicone content is high enough to improve oxygen permeability, the surface characteristics of the contact lens become hydrophobic and cannot be worn without surface modification to make the surface at least moderately hydrophilic to allow contact with tear fluid and epithelium of the cornea. All silicone/hydrogel contact lenses with high enough oxygen permeability have some kind of surface modification, as seen in Table 10.5. Then, how to make the hydrophobic surface of a silicone polymer phase hydrophilic and how durable is the modified surface become key factors that determine the effectiveness of silicone/hydrogel contact lenses.

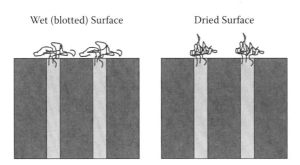

FIGURE 10.13
An LBL coated silicone/hydrogel contact lens surface.

A silicone/hydrogel contact lens surface is composed of a hydrophilic water-swollen hydrogel phase and a hydrophobic dry silicone phase. It is necessary to convert the hydrophobic dry silicone polymer phase to a hydrophilic water-wettable surface without drastically reducing the oxygen permeability of the silicone polymer phase. Conventional wet chemistry processes are practically useless, because the chemicals that make the surface wettable do not wet the hydrophobic silicone polymer surface. The wet chemical process known as the layer-by-layer (LBL) process could be applied to a water-equilibrated silicone/hydrogel contact lens surface, as mentioned in Chapter 3, but the coating does not adhere to the silicone polymer phase and adds on to the hydrogel phase as depicted in Figure 10.13. In water, the extended polymer segments of hydrophilic polymers might extend to partially cover the surface of the silicone polymer phase, but without adhering to the surface of the hydrophobic phase. Such a coated surface, however, is highly susceptible to the change of the surrounding medium (e.g., blotting of surface, drying, etc.). It is important to recognize that the surface is hydrophilic only when liquid water is held on it, and the removal of liquid water from the surface turns the surface hydrophobic (contact lens greater and 90°), except for that with plasma polymerization coating.

Plasma polymerization coating is a dry process applied on dry surfaces; consequently, the coating applies equally to both the silicone polymer phase and the hydrophilic polymer phase, as schematically depicted in Figure 10.14. However, the coated silicone/hydrogel contact lens is rehydrated to make hydrogel phase hydrated, and the plasma polymerized nanofilm might cause nanoscale to microscale cracking depending on the hydration level of the hydrogel phase. In general, the higher the hydrogel content, the higher is the trend of micro cracking, which does not change the hydrophilicity of the lens but could reduce the resistance for penetration of lipid-soluble substances (i.e., the transport cut-off point increases appreciably but not at a catastrophic level). With a silicone/hydrogel contact lens that has higher

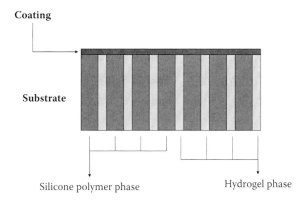

FIGURE 10.14
Plasma polymerization coating on a phase-separated silicone/hydrogel contact lens surface.

silicone content and a lower hydrogel component, nanocracking does not influence the transport characteristics.

The addition of a hydrogel phase makes the processing of dry plasma polymerization coating significantly cumbersome, because most of the water in the hydrogel phase should be removed before the dry processing is performed. The removal of water from the hydrogel phase requires more time than one might imagine. If one tries to remove water quickly, the fast removal from the top layer of the surface forms a dry skin that slows the drying of water from subsequent layers of a film. As mentioned in Chapter 8 (Figure 8.1), the level of vacuum or low system pressure cannot be used to judge the extent of the removal of water from the substrates to be coated.

The free volume of silicone polymer is an advantageous feature for high oxygen permeability, but an undesirable feature for absorbing lipid-soluble substances from the biological system in contact. For temporary use of contact lenses (e.g., for daily disposable contact lenses) and for daily use for a short time (e.g., for a week), the problem would not be recognized; however, it is not a desirable feature, and the coating that prevents it is preferred. For extended-wear contact lenses, it becomes an unacceptable feature of silicone/hydrogel contact lenses.

The high oxygen permeability, on the other hand, does not seem to play a key role in short-time wearing, because the time lag of oxygen diffusion through a contact lens is in the order of a minute, while the frequency of a human blinking an eye is in the order of 5 seconds, which means that fresh oxygen-saturated tear fluid is provided to the cornea every 5 seconds. This situation changes completely in the case of extended-wear contact lenses. In that case, oxygen solubility in the silicone phase of a silicone/hydrogel contact lens, rather than high oxygen permeability, seems to play the key role [19].

The requirements for coating silicone/hydrogel contact lenses can be summarized as follows:

1. The coating should be applicable to the hydrophobic silicone polymer phase and impart a hydrophilic surface.
2. The coating should not reduce oxygen permeability of the silicone polymer phase at an excessive level.
3. The coating should work as a barrier to lipid-soluble substances.
4. The coating thickness should not exceed ca 40 nm in order to prevent two layers (two bulk phases) of polymer film (i.e., stay within the thickness of surface state).

It is clear that MLCVD CH_4 nanofilm is an ideal process to fulfill the above requirements. In the new revelation of unique dielectric breakdown of the gas phase in low pressure under the interactive influence of a magnetic field, no other method of coating could compete with MLCVD amorphous carbon nanofilm coating in the area of surface preparation for biocompatibility.

10.2.3.2 Industrial-Scale Batch and Continuous Operation of MLCVD CH_4 Coating

Production of large numbers of coated contact lenses can be done by batch operation or continuous operation depending on the quantity of products. In batch operation, multiplicities of the reactor used in the laboratory-scale exploitation of the method are used as depicted in Figure 10.15 [39]. Because of unique MLCVD reaction mechanisms (described in Chapter 9), successive batch operation can be carried out for a month before a preventive maintenance break. Two operators could maintain programmed successive batch operation, except mounting of contact lenses on a sample holder. Up to 20 million contact lenses are probably produced each year, and this mode of operation would suffice.

For production of larger quantities, the reactor for continuous operation would be advantageous. Figure 10.16 [39] depicts a typical continuous mode reactor. The most critically important and labor-intensive operation in both modes of operation is the setting up of sample holders, for which the extent of labor intensity is nearly the same. Figure 10.17 [39] shows the sample holder with contact lenses which is used in a bell-jar-type reactor. Figure 10.18 [39] shows, for the continuous feeding reactor, which flow sheet of processing is depicted in Figure 3.2 (Chapter 3).

In both modes of operation, substrates move in and out of the steady-state luminous gas phase established in between magneto-electrodes. The motion of substrate in the batch operation and in the continuous operation is schematically depicted in Figures 10.19 and 10.20, respectively. Because

FIGURE 10.15
Successive batch operation.

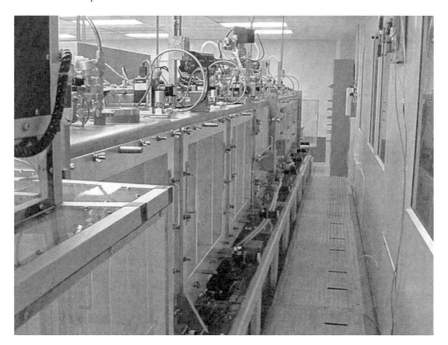

FIGURE 10.16
Continuous linear motion reactor.

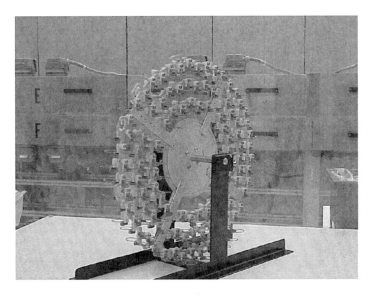

FIGURE 10.17
Sample holder for batch operation.

FIGURE 10.18
Sample holder for linear motion continuous reactor.

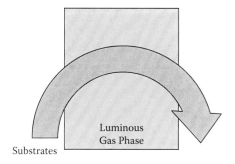

FIGURE 10.19
Substrate movement with respect to a steady-state luminous gas phase in the batch reactor.

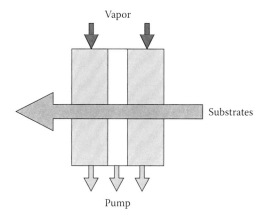

FIGURE 10.20
Substrate movement with respect to the steady-state luminous gas phase in a continuous linear motion reactor.

of the toroidal glow, the movement of substrates in the luminous gas phase is mandatory, which yields uniform distribution of deposition within a substrate as well as uniformity among contact lenses treated within a sample holder.

10.2.4 Encapsulation of Metallic Stent

When the MLCVD of CH_4 coating was applied on the surface of memory recovery expandable stainless steel stents (Walstent), of which the bare surface has poor blood compatibility, and implanted in a pig without using any drugs to suppress blood coagulation, all five coated samples stayed patent, while uncoated stents with drug showed partial to total closure [23] as shown in Figure 10.21. The authors of the paper on in vivo experiments were not

• **Pig model experiments:** open/closed

    – Uncoated Stent   **0/5**

    – Uncoated Stent with coagulation depressing drug
        in blood   **1/4**

    – MLCVD CH$_4$ coated stent without drug in blood
                                                        5/0

FIGURE 10.21
Pig model experiment results of MLCVD CH$_4$ coated stents.

aware of the details of the coating and reported it as a "polymer coated stent." Such a clear-cut result (i.e., patency 5 out of 5), has never been seen in any animal experiments with artificial material, which again seems to indicate the superior biocompatibility of the imperturbable surface of the MLCVD CH$_4$ coating. The stent was licensed to a leading pharmacology/biomedical device company that decided prior to the license agreement that they would be out of cardiovascular applications of their products; hence the MLCVD CH$_4$ coated stent has never been used in cardiovascular applications.

One concern during the period of coating surfaces of the stent was if it was possible to coat the inside lumen of a metallic stent, because the stent would be a typical "Faraday Cage." It turned out that there is no problem coating all surfaces of the metal, which confirmed our view that what is generally recognized as ionized gas (low-pressure plasma phase) is a virtually neutral luminous gas phase. The metallic stent, which could be a Faraday Cage, moves into the luminous gas phase (i.e., it is not attempted to create ionized gas within a Faraday Cage).

All characteristics of MLCVD CH$_4$ coating described in the contact lens applications also apply to metallic stent coating. The application of the MLCVD CH$_4$ coating on metal substrate is much simpler because the pretreatment or handling of substrate prior to the coating applications is much simpler and easier. The stent coating was investigated prior to the coating of silicone/hydrogel contact lenses.

10.2.5 Plasma System Approach Interface Engineering for Biomedical Electronics Devices

Many electronic devices used in biomedical applications, such as electrical pulsing and stimulation, and measurement of electric current for diagnostic data collection, require use of a noble metal electrode, such as platinum or gold, because of the biocompatibility required for the devices. The inertness of metal is vitally important for the biocompatibility. On the other hand, the insulation of noble metal, particularly when insulated metal is used in contact with liquid water or body fluid, is difficult, because most insulating materials do not adhere well to the surfaces of noble metal because of its inertness.

Furthermore, such a device requires pinpoint electron flux at the designated open surface of the metal, which means that the interface of the insulation material to the noble metal must be exposed to the biological system. If we could not maintain the exposed interface intact, and water penetrates along the interface, the pinpoint flux of electrons gradually shifts to a massive leak of electrons. Unless this characteristic problem with noble metals can be solved, such devices could cause catastrophic failure. This problem also points out that we should know how insulation fails. The concept of "salt intrusion" rather than "salt diffusion" described in the preceding section provides a guideline for how to solve the problem. One requirement is the use of a low-crystallinity polymer or a totally amorphous polymer, of which successful utilization requires the knowledge of plasma system approach interface engineering (P-SAIE). Some examples of the importance of P-SAIE are discussed in the following sections.

Electrochemical impedance spectroscopy (EIS) is a valuable method to study the barrier properties and corrosion protection performance of polymer-coated metals. It has been widely used in this field in recent years [24–28]. Many examples can be found in the literature, which illustrate the performance deterioration of different coatings on metals as well as pretreatment effects on the properties of a coating system [26–28]. EIS can also be applied to a freestanding film; however, only a few studies with this use of EIS have appeared in the literature prior to the study summarized below [29–31].

The principle of EIS is schematically depicted in Figure 10.22. The resistivity as a function of the frequency of applied electrical power is taken

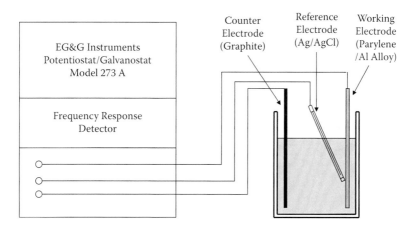

FIGURE 10.22
EIS for insulation characteristics study.

periodically (Bode plots). Bode plots for a fresh sample usually show a straight line that is the reference line for investigation of ion-transport characteristics of the coated system. If an interface engineered system is good (i.e., resistivity is not influenced by immersion in a buffered salt solution), the initially found straight line does not change with immersion time. The deterioration of the insulation characteristics of sample shows first the lowering of resistivity in a low-frequency range (i.e., the bending of the initial straight line to lower resistivity could be taken as a sign of deteriorating insulation, and progresses further with time).

The overall corrosion protection of a metal, as well as general protective coatings irrespective of the function of the coatings, depends on the performance of a system as a whole, including many interfaces and coating layers. These factors are not only mutually dependent as a combination, but also in the order of application (permutation). Any single factor cannot be treated as a dominant one. There is very little work in the literature that focuses on the role of interfacial factors in the corrosion protection of coated systems.

An ideal model system was selected to study the interfacial factors with EIS [32]. The model system was Parylene C coated Alclad (aluminum clad aluminum alloy), which is schematically depicted in Figure 10.23. In this system, the

Combination of Interfacial Factors for Each Sample Studied

Sample No. & Top Plasma Polymer	Adhesion of Parylene C to Metal	Liquid-contacting Surface	Polymer Contacting Metal
(1) None	No	PC	PC
(2) None	Yes	PC	PP
(3) Hydrophilic	No	Plasma Polymer of TMS/O$_2$	PC
(4) Hydrophilic	Yes	Plasma Polymer of TMS/O$_2$	PP
(5) Hydrophobic	No	Plasma Polymer of TMS	PC
(6) Hydrophobic	Yes	Plasma Polymer of TMS	PP

PC: Parylene C,   PP: Plasma Polymer

FIGURE 10.23
Factors involved in Parylene C coated aluminum sheet.

surface state of the top surface (salt solution/coating interface) and the adhesion of the coating (coating/metal interface) were modified to study the influence of these factors on the corrosion protection performance of the system.

Parylene C film does not adhere to any smooth surface due to its unique polymerization mechanisms. Parylene polymerization, in which the di-radicals Parylene react only with free radicals (i.e., no free radicals on the surface, no adhesion of Parylene film). Adhesion can be improved to an excellent level by means of plasma polymerization of an ultra-thin layer to which a Parylene C film will bond covalently.

A freestanding Parylene C film can be easily peeled off of the substrate surface, although a film does not peel off by itself in many cases, even if immersed in water. This feature enables us to investigate a system with poor adhesion, and also to investigate a freestanding film with EIS. Figure 10.24, which depicts Bode plots of freestanding Parylene C film, shows that Parylene C film is a poor insulator in an aqueous salt solution; the initial line is bending at the low resistivity level and declines further with time. The poor insulation is likely due to the salt intrusion described in Section 10.3.1, because Parylene C film is semicrystalline. Examination of these three cases—a freestanding film, a system with poor adhesion, and a system with good adhesion—made it possible to see how EIS data would change when salt solution penetrates the interface of coating and substrate metal.

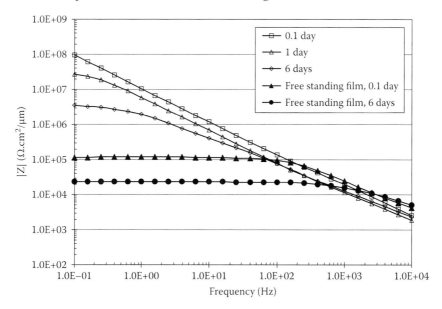

FIGURE 10.24

Changes of Bode plots as a function of immersion time for freestanding Parylene C film and Parylene C coated Alclad 7075-T6 aluminum sheets in 0.9% NaCl solution. 0.1 day indicates the initial run after 2 hours immersion of the samples.

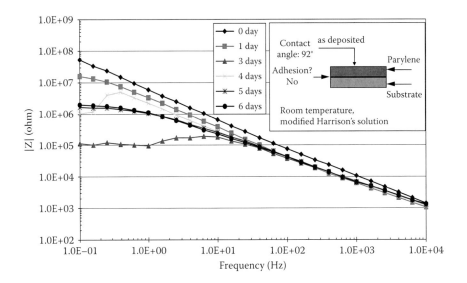

FIGURE 10.25
EIS Bode plots for Parylene C coated sample without pretreatment and surface treatment, Case (1).

EIS data obtained for these samples are shown in Figures 10.25 through 10.30. These figures, with explanations of interfacial factors for each sample, provide step-by-step illustrations of factors involved in the interface engineering processes. The data revealed the following important aspects that have significant implications in the corrosion protection provided by a coating.

1. Parylene C, which has excellent barrier characteristics, did not perform well in corrosion protection. The salt intrusion resistance is poor, probably due to its semicrystalline nature of the bulk phase. The boundary phase between the crystalline and amorphous phase seems to be vulnerable for salt intrusion.

2. The change in Bode plot as a function of immersion time for a Parylene C coated metal sheet is greatly influenced by the nature of the top surface (solution–coating interface).

3. The hydrophilic top surface accelerates time-dependent change, while the hydrophobic top surface decreases the extent of the change.

4. The change in Bode plot as a function of immersion time for a Parylene C coated metal sheet is greatly influenced by the degree of adhesion of Parylene C to the metal substrate surface.

5. Without good adhesion, time-dependent change Bode plots converge to that for a freestanding film, which is strong evidence for interfacial failure.

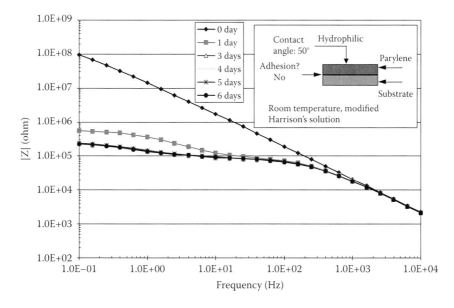

FIGURE 10.26
EIS Bode plots for Parylene C coated sample with the pretreatment to impart adhesion but no surface modification, Case (2).

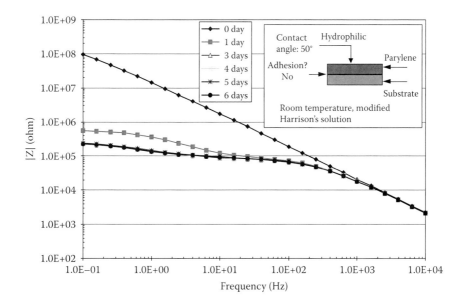

FIGURE 10.27
EIS Bode plots for Parylene C coated sample without pretreatment but with surface modification to make the surface hydrophilic. Case (3).

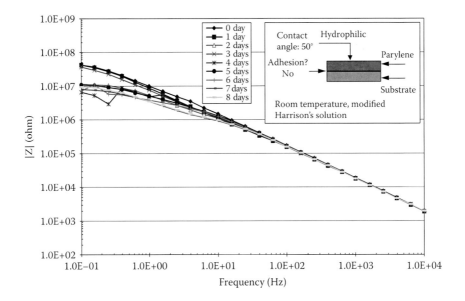

FIGURE 10.28
EIS Bode plots for Parylene C coated sample with pretreatment and the surface modification to make the surface hydrophilic, Case (4).

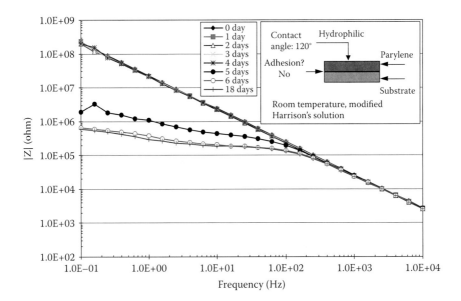

FIGURE 10.29
EIS Bode plots for Parylene C coated sample without pretreatment but with the surface modification to make the surface hydrophobic, Case (5).

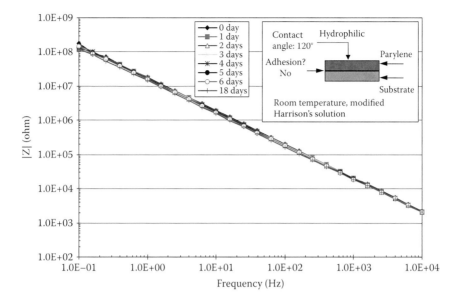

FIGURE 10.30
EIS Bode plots for Parylene C coated sample with pretreatment and the surface modification to make the surface hydrophobic, Case (6).

6. Among the factors investigated, the adhesion of polymer film to the substrate metal plays the most significant role in improving the overall corrosion protection of the system; however, this factor alone is not enough to yield the best corrosion protection.

7. With a hydrophobic plasma polymer on top and good adhesion to the substrate metal, the Bode plot did not change with immersion time. Only with the optimization of all of these factors was the excellent corrosion protection by Parylene C coating obtained.

Thus, without P-SAIE, Parylene C film, which has excellent barrier and physical properties, cannot be utilized in corrosion protection of a metal. Conversely, P-SAIE is the key to yielding excellent corrosion protecting systems. It is also important to recognize how a nanofilm of hydrophobic amorphous network of plasma coating can prevent the initiation of the salt intrusion process.

The incorporation of electrochemical corrosion inhibitors is the current mainstream approach for corrosion protection, but this cannot be utilized in biomedical applications. The "water-insensitive adhesion" of insulating material to a noble metal surface is the key principle. Furthermore, they are environmentally benign and free from the health hazards associated with conventional methods. The nature of the top surface, bulk phase, and the interface to the substrate are the key factors of P-SAIE, which are also applicable to any functional coating to protect the substrate.

Results clearly show that P-SAIE is necessary to obtain excellent insulation of a metallic conductor used in microelectronic systems to be used in biomedical applications. The only acceptable result shown in Figure 10.30 is the product of P-SAIE, not the consequence of combining two good materials together. The best result obtained with the P-SAIE Parylene system showed no deterioration of the EIS test in 18 days. How such results correlate to real cases is the subject of future studies; however, prolonged testing is meaningful only if systems show excellent results in the short-term (18 days) screening tests. The data shown above clearly demonstrate the merit of P-SAIE. The application of what we found with macrosystems to microsystems is one key area where MLCVD (LCVD with magnetic field) would play an important role.

10.2.6 Unique Features of MLCVD Amorphous Carbon Nanofilm for Biocompatibility

Common denominator factors found by applying nanofilm of amorphous carbon prepared by MLCVD could be summarized as follows. Some features listed could be restatements of other listed features, but they nevertheless deserve attention. The most important point here is that all features described are integrated into the nanofilm regardless of the nature of the substrate materials.

1. The surface is imperturbable by the contacting medium at the interface. The contact angle of water does not change regardless of change of surrounding medium: liquid water, blotted wet surface, dry in air, and so forth.

2. The adhesion of the nanofilm to the substrates, which could pass qualifications to be used in biomedical applications, is excellent. If it is necessary, adhesion could be further improved by applying plasma system approach interface engineering.

3. The surface can be autoclaved, repeatedly if necessary. Surface characteristics do not change after 10 times of autoclaving coated silicone/hydrogel contact lens. This is a very important feature that any artificial materials that are to be used in biomedical applications should have.

4. The surface is robust and resists most ordinary modes of surface rubbing considered in the evaluation of the durability of an artificial material in biomedical applications.

5. Protein adsorption is very low as shown in Figure 10.31 [40], which indicates that adsorption of protein is less than that on the gold surface used as the reference surface of the test method. Furthermore,

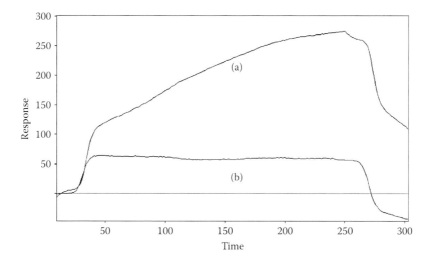

FIGURE 10.31

Time-dependent adsorption of protein, bovine serum albumin, in an SPR measurement setup from 10 µg/mL PBS (pH 7.4) standard phosphate buffer solutions, at a flow rate of 50 µL/min; on (a) an uncoated gold surface (standard reference), and (b) an MLCVD coated reference surface. The experiment was stopped after 250 min. To wash the whole system, the flow rate was increased to 250 µL/min.

the adsorbed protein thickness starts to decrease as soon as the washing cycle of the testing procedure starts, indicating no adhesion of protein occurs. Likewise, lipid and lipid-soluble substances cannot adhere to the surface.

6. Bacteria adsorption on the surface is slower and reaches the lower plateau value than those for the gold surface as depicted in Figure 10.32 [40], as a comparative observation of bacterial adsorption on gold reference surface and on the coated gold reference surface. The same trend is found with protein adsorption. The tests of adhesion characteristics with higher flow rate of washing cycle need further evaluation.

7. The nanofilm does not allow penetration of lipids and lipid-soluble substances through the film, and effectively prevents the swelling of the silicone polymer phase by lipids.

8. The nanofilm, with thickness ca 20 nm, is flexible enough to allow flexing of the coated contact lens without causing delamination or cracking of the nanofilm.

9. Scrupulous cleaning of the surface is unnecessary after use in contact with biological systems.

10. The surface allows intimate contact with biofluid (e.g., tear film), tissue cells, and so forth, without adhering or sticking.

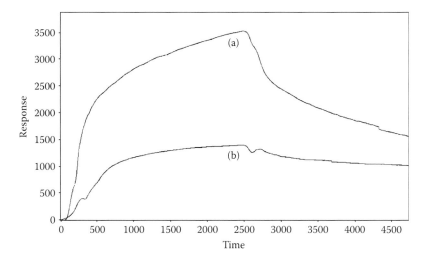

FIGURE 10.32
Time-dependent, unspecific bacterial adsorption was measured in a SPR measurement setup with 1×10^8 *Enteroccocus faecalis* in LB-media with a flow rate of 50 µL/min: (a) uncoated gold surface (standard reference), and (b) MLCVD coated reference surface; the experiment was stopped after 250 min, but in this case, the flow rate was not increased.

10.3 Interface Engineering for Adhesion of Coating

10.3.1 Salt Diffusion versus Salt Intrusion

The most damaging factor of corrosion of metals can be attributed to the poor adhesion in the presence and influence of aqueous ionized salt solution at the interface of coating and metal, which is absent in a freshly prepared coated metal. The obvious question that arises is how salt gets into the interface: the mode of transport of salt through the coating polymer layer. Another problem, which is a persistent problem with microsensors and microstimulating devices, which utilize electrical current, is how we could prevent the lateral advance of salt solution along the interface of the electrical insulator and conductor.

Without understanding the fundamental mode of salt transport in polymeric dielectric materials, those crucially important failures could not be prevented. The data shown in the following section are obtained with the cathodic plasma polymerization, Cathodic LCVD, dealing with larger metal plates. Dealing with smaller and more intricate substrate (e.g., microsensor, microelectrical stimulator, etc.) for biomedical applications, MLCVD could be utilized more effectively, handling a larger number of substrates at one processing.

The surface state of a semicrystalline polymer, according to the three-phase model of semicrystalline polymers, consists of the crystalline phase, the amorphous (noncrystalline) phase, and the transition phase between the crystalline and the amorphous phases. Most physical properties, including the transport of small molecules through the polymer phase, have been treated according to the two-phase model: crystalline and amorphous phases. However, for the adequate interpretation of the topics addressed, it is necessary to adopt the three-phase model of semicrystalline polymers.

The macromolecules in the crystalline phase are considered to be immobile, and it is considered in all practical sense as impermeable for most permeants. The macromolecules in the amorphous phase are mobile, of which mobility depends on $(T - T_g)$, and the phase is considered to be permeable (i.e., the permeability is dependent on the volume fraction of the amorphous phase). The macromolecules in the transitional phase are tie molecules, of which a part is in the crystalline phase and another part is in the amorphous phase; the mobility is restricted to in between those of crystalline and amorphous phases.

The transitional phase does not extend far from the surfaces of crystals, but the volume fraction of the transitional phase, though it is small, is proportional to the volume fraction of the crystalline phase. Thus, the higher is the crystallinity, the higher is the effect of the transitional phase. The role of the transitional phase seems to be very important in determining the stability of transport resistance of an insulator, because the deterioration of the resistivity of an insulating layer does not follow the gradual decay due to the change of the diffusion transport characteristics of the amorphous phase, but it occurs as a sudden abrupt change as described below.

Hydrated ions are much larger than water, and hydrated cation and hydrated anion must move together because of the Columbic attractive force between them. Consequently, salt ions cannot permeate through an amphoteric hydrophobic/hydrophilic polymer, of which the hydration value is low (i.e., less than a few volume percent, by the solution-diffusion principle). Therefore, salt permeation through a hydrophobic polymer film, such as low-density polyethylene (LDPE) and high-density polyethylene (HDPE), used for insulation of the distribution power cable insulation should not occur.

In reality, however, salts dissolved in water find or create paths into and through a hydrophobic polymer matrix and cause the breakdown of an insulating layer or the corrosion of the substrate metal. In contrast to the diffusion process, the process of salt going into the polymer matrix could be termed as *salt intrusion*, because the breakdown of the surface state does not occur with water that does not contain salts. The salt intrusion starts with the breakdown of the surface state of an insulator in contact with a salt solution. In a study of the electric insulation characteristics of LDPE film, it was found that salt ions intrude into the polymer matrix by salt intrusion mechanisms [21,22]. The phenomenological salt intrusion found can be summarized as follows:

1. The alternating current (AC) resistivity of LDPE film does not change with water immersion time. Figure 10.33 depicts the change

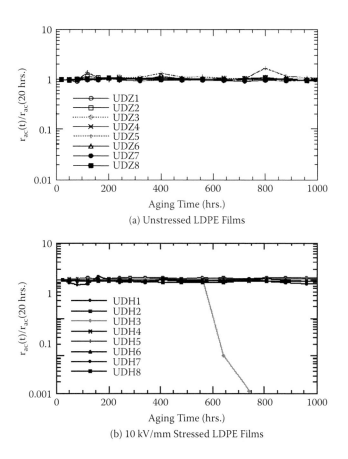

FIGURE 10.33
AC resistivity ratio versus aging time for untreated LDPE films in a de-ionized water environment: (a) unstressed and (b) 10 kV/mm stressed.

of relative resistivity with immersion time with no electrical stress and with electrical stress (10 kV/mm). These figures clearly show that water does not cause the insulation failure.

2. When an LDPE film is immersed in a salt solution (0.9% NaCl), the AC resistivity decreases as a function of the immersion time, as shown in Figure 10.34. These figures include the effect of a nanofilm of plasma polymer deposited on the surface of LDPE. With hydrophobic plasma polymer (HFE + H_2), the decrease of AC resistivity was not observed. These figures indicate that the surface-state breakdown occurs when the salt intrusion takes place. The salt intrusion can be prevented by the application of a plasma polymer, which is an amorphous network (one phase, and no weak boundary). The extent of protection seems to be dependent on the hydrophobicity of the network.

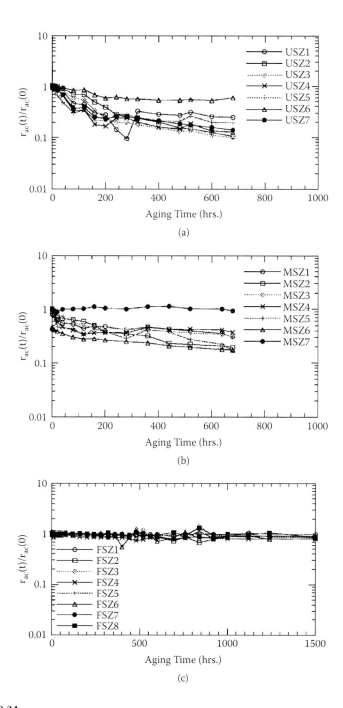

FIGURE 10.34

AC resistivity ratio versus aging time for unstressed LDPE films in a 0.9% saline environment: (a) untreated, (b) CH_4 plasma treated, and (c) $C_2F_6 + H_2$ (1:1) plasma treated.

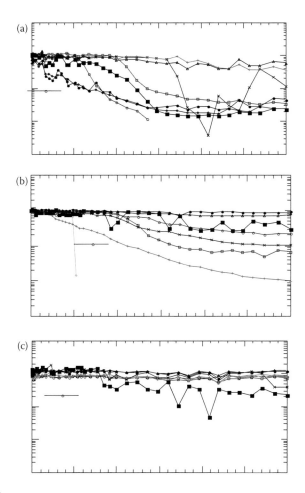

FIGURE 10.35
AC resistivity ratio versus aging time for 5 KV/mm stressed samples in a 0.9% saline solution:
(a) untreated, (b) CH_4 treated, and (c) $C_2F_6 + H_2$ (1:1) treated.

3. The insulation breakdown under electrical stress occurs in a fatigue
   mode (i.e., not a gradual deterioration but an abrupt failure) and is
   correlated with the salt intrusion characteristics of the film, as shown
   in Figures 10.34 and 10.35. The hydrophobic surface state created by
   plasma polymerization of (HFE + H_2) significantly prolongs the time
   when the breakdown of surface state by salt intrusion occurs.

Based on these observations, the distinction between water penetration and
salt intrusion may be schematically represented as shown in Figures 10.36
and 10.37. In these figures, there are two interfaces (i.e., film–water and film–
metal) involved, and the effect of lack of water-insensitive adhesion at the

Water Diffusion Through Coatings

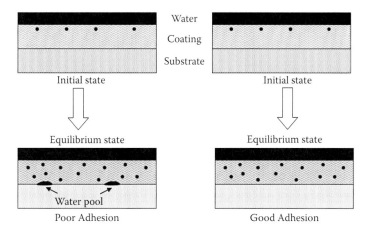

FIGURE 10.36
Water diffusion through a coating with and without water-insensitive adhesion of the coating to the substrate; salt diffusion rides on the diffusion of water, but slower than water without electrical stress.

Salt Intrusion Through Coatings

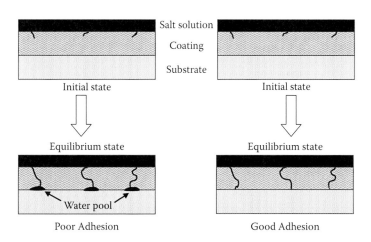

FIGURE 10.37
Salt intrusion through a coating with and without water-insensitive adhesion of the coating to the substrate.

film–metal interface is also depicted. In the absence of water-insensitive adhesion, water molecules that reach the polymer–metal interface cluster together by breaking weak polymer–metal interactions. If one assumes the diffusion constant of water through a polymer to be 10^{-8} cm^2/s, the time lag of water diffusion through 15 μm thick LDPE film is estimated to be 38 s. Any indication of the salt intrusion effect appears in a much longer period of time (i.e., days and months). Therefore, it can be assumed that the intrusion occurs in a water-saturated polymer matrix. It is important to recognize that the surface state breakdown at the interface of liquid–top layer of insulator is the initial step of salt intrusion. The effective prevention of salt intrusion can be achieved by placing totally amorphous plasma polymer on the surface of the insulator (LDPE), which also supports the concept that the transitional phase (between crystalline and amorphous phases) is responsible for the surface-state breakdown.

10.3.2 System Approach Interface Engineering (SAIE)

10.3.2.1 Why Do We Need System Approach Interface Engineering (SAIE)?

System Approach Interface Engineering (SAIE) is the tailoring of interfaces to fit the requirements for the multilayer system to accomplish the objectives. As described before, it seldom occurs that the bulk property of a material and the surface characteristics match the requirements for use of the material in a specific purpose. Consequently, it is nearly imperative to cope with the requirements for the bulk properties and the surface characteristics separately, and then combine them together as the final product that satisfies the overall requirements. A coated metal is an example of the system that is designed to protect the metal from the corrosion. The objective of the system is to protect the metal, and the coated metal is the total system. Although it could be conceived as a simple metal coated with a paint (i.e., two phases consisting with a metal and a coating), it generally requires multiple layers of coating. Such a seemingly simple system actually involves many important interfaces, if one looks at such a simple system from the viewpoint of how the surface state of the two bulk phase changes at the transitional zones of each material: corrosion-protecting layer, primer for adhesion, main coating, top surface coating, and so forth.

The modern coatings of metals for corrosion protection are actually practiced by the fundamental principle of SAIE; however, it is carried out mainly with wet chemistry of galvanic corrosion protection. The persistent problem is the fact many corrosion-protecting chemicals, such as chromium-containing chemicals, are environmental hazards in the application process as well as in the recycling of metals. Only "rear-end approach" environmental mediation can be applied to take care of environmentally hazardous materials, which practically means the problem would persist and hazardous materials would accumulate in our environment.

10.3.2.2 Conventional SAIE and Plasma SAIE (P-SAIE)

The significance of Plasma System Approach Interface Engineering (P-SAIE) is that it is a "front-end" approach green process, which does not use potentially hazardous material to protect corrosion. P-SAIE could be explained by examples of corrosion protection of aluminum alloys by applying a protective coating [26–29]. Figure 10.38 depicts the interface between a coating and a metal substrate, in which the modification of oxides, application of an ultra-thin layer of plasma polymerization coating, and application of 10 to 30 μm thick primer layer are involved. With consideration of the boundary of oxides and the metal bulk phase, which is not indicated in the figure, there are 10 interfaces and boundaries to be considered.

The surface of a pure aluminum sheet or film is covered by aluminum oxides, which are stable and provide excellent corrosion protection of the pure aluminum under normal environmental conditions. Aluminum corrodes in high (greater than 11.5) and low (less than 2) pH solutions with specific exceptions like strong acid solutions of high reduction/oxidation (Redox) potential (e.g., concentrated nitric acid in which aluminum is passivated). However, pure aluminum is too soft, and the mechanical strength is not sufficient for many practical uses. Alloying provides improved mechanical strength and properties, but at the expense of corrosion resistance, because oxides are mixed oxides according to the content of alloying elements, which are not as stable as aluminum oxides and are vulnerable to corrosion. The surface oxides on Al alloys offer relatively good corrosion protection in mildly corrosive environments, especially in comparison to the case of steel; however, further protection of surface oxides is necessary.

A corrosion protection system should include protection of the oxides and, in addition, should provide a good adhesive base for subsequent

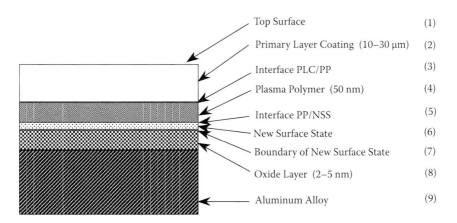

FIGURE 10.38
The SAIE system.

coating. The conventional corrosion protection system consists of alkaline cleaning and deoxidization of the surface followed by the application of a chromate conversion coating. The purpose of pretreatments is to remove the surface contaminants and, thus, create a clean surface on which chromium oxide can be grown, which then acts as the corrosion protective layer and the adhesive base. The conventional corrosion-protecting system with chromate conversion coating can be visualized by replacing the plasma polymer layer in Figure 10.38 with a chromate conversion coating. The P-SAIE approach in the corrosion protection has an important advantage of being able to eliminate chromate conversion coating, which is an environmental and health hazard, and achieve the same or better corrosion protection.

According to the principle of SAIE, the combination of the best elementary factors (e.g., the best primer and the best paint) does not necessarily yield the best performance of the overall coated system—the painted subject. The change of one component (e.g., primer coating) may necessitate the change of a pretreatment procedure. The best result out of the change of primer could be obtained after all other factors are optimized to the new primer. This aspect could be visualized by the following examples.

The electrolytic deposition of a coating that is known as "E-coat" provides an excellent corrosion protection of steel. Today, nearly all automobiles are corrosion protected by applying the cathodic E-coat, in which the steel body of a car is used as the cathode of the electrolytic deposition of a primer coat, on the surface of zinc-phosphated steel. It is quite logical to imagine that if an E-coat is applied to chromate conversion coated aluminum alloy surface, a significant improvement of the corrosion protection of aluminum alloys could be realized, because such an attempt represents the combination of the two best components (i.e., chromate conversion coating and E-coat). In reality, the combination of the two best coatings is not the best solution to produce the best result, but the best solution lies in the adoption of SAIE.

P-SAIE is an advanced SAIE process, which could be green process if the selection of plasma process is carried out with consideration of green aspects. Magneto-luminous chemical vapor deposition (magneto-plasma polymerization) is a super-green process, which is ideally suited for bio-medical applications. MLCVD-SAIE could revolutionize bio-materials, which are used in contact with biological systems in vitro, in ex-vivo, and in vivo. The crucially important factor, however, is that it should be utilized with serious efforts to understand the underlying principle of each step; it is not an almighty process to be used with wish for a quick success. The true value of MLCVD would be realized with future studies on the fundamental process and utilization in selected applications.

References

1. Yasuda, H., *Luminous Chemical Vapor Deposition and Interface Engineering*, CRC Press, Boca Raton, FL, 2004.
2. Lin, J. N., S. K. Banerji, and H. Yasuda, *Langmuir*, 10, 936, 1994.
3. Lin, J. N., S. K. Banerji, and H. Yasuda, *Langmuir*, 10, 945, 1994.
4. Holly, F. J. and M. F. Refojo, *Journal of Biomedical Materials Research*, 9, 315, 1975.
5. Baskin, A., M. Nishino, and L. TerMinassian-Sarage, *Journal of Colloid and Interface Science*, 54, 317, 1976.
6. Ratner, B., P. Weatherby, A. Hoffman, M. Kelly, and L. Scharpen, *Journal of Applied Polymer Science*, 22, 643, 1978.
7. Thomas, R. and R. Trifilet, *Macromolecules*, 13, 45, 1979.
8. Sung, C. and C. Hu, *Journal of Biomedical Materials Research*, 13, 45, 1979.
9. Yasuda, H., A. K. Sharma, and T. Yasuda, *Journal of Polymer Science: Polymer Physics Edition*, 19, 1285, 1981.
10. Hanazawa, E. and R. Ishimoto, *Nippon Sechaku Kyokaishi*, 18, 247, 1982.
11. Hanazawa, E. and R. Ishimoto, *Nippon Sechaku Kyokaishi*, 19, 95, 1983.
12. Harttig, H. and K. Huttinger, *Journal of Colloid Interface Science*, 48, 520, 1983.
13. Yasuda, T., T. Okuno, H. Yoshida, and H. Yasuda, *Journal of Polymer Science Part B: Polymer Physics*, 26, 1781, 1988.
14. Yasuda, T., H. Yoshida, T. Okuno, and H. Yasuda, *Journal of Polymer Science Part B: Polymer Physics*, 26, 2061, 1988.
15. Yasuda, H., E. J. Charlson, E. M. Charlson, T. Yasuda, M. Miyama, and T. Okuno, *Langmuir*, 7, 2394, (1991).
16. Yasuda, T., T. Okuno, and H. Yasuda, *Langmuir*, 10, 2435, 1994.
17. Yasuda, H., T. Okuno, Y. Sawa, and T. Yasuda, *Langmuir*, 11, 3255, 1995.
18. Langmuir, I., *Science*, 87, 493, 1938.
19. Yasuda, H., *Macromolecular Bioscience*, 6, 121, (2006).
20. Refojo, M. F. and F. L. Leong, *Contact and Intraocular Lens Medical Journal*, 7, 226, 1981.
21. Ho, C.-P. and H. Yasuda, *Journal of Biomedical Materials Research*, 22, 919, 1988.
22. Ho, C.-P. and H. Yasuda, *Journal of Applied Polymer Science*, 38, 741, 1989.
23. van der Giessen, W. J., H. M. M. van Beusekom, C. D. van Houten, L. J. van Woerkens, P. D. Verdouw, and P. W. Serruys, *Coronary Artery Diseases*, 3, 631, 1992.
24. van Westing, E. P. M., G. M. Ferrari, and J. H. W. De Wit, *Corrosion Science* 36(8), 1323, 1994.
25. Amirudin, A. and D. Thierry, *British Corrosion Journal*, 30(2), 128, 1995.
26. Cohen, S. M., *Journal of Coatings Technology*, 68 (859), 73, 1996.
27. Tang, N., W. J. van Ooij, and G. Górecki, *Progress in Organic Coatings*, 30, 255, 1997.
28. Mertens, S. F., C. Xhoffer, B. C. De Cooman, and E. Temmerman, *Corrosion*, 53(5), 381, 1997.
29. Barreau, C., D. Massinon, and D. Thierry, *SAE Transactions*, 100, 1281, 1991.
30. Sussex, G. A. M. and J. D. Scantleburry, *Journal of the Oil and Colour Chemists' Association*, 66, 142, 1983.
31. Walter, G. W., *Corrosion Science*, 32, 1085, 1991.

32. Yasuda, H., Q. S. Yu, and M. Chen, *Progress in Organic Coatings*, 41, 273, 2001.
33. Yasuda, H., E. J. Charlson, E. M. Charlson, EPRI Report, "Effect of Plasma Polymer Coating on the Insulation Breakdown," 1995.
34. S. Lee, Y., dissertation, University of Missouri–Columbia, "Effects of Plasma Polymer on the Multi-Stress Aging of Organic Insulation and Proposed Degradation Mechanisms," 1995.
35. Reddy, C. M., Q. S. Yu, C. E. Moffitt, D. M. Wieliczka, R. Johnson, J. E. Deffeyes, and H. K. Yasuda, *Corrosion*, 56, 819, 2000.
36. Yu, Q. S., C. M. Reddy, C. E. Moffitt, D. M. Wieliczka, R. Johnson, J. E. Deffeyes, and H. K. Yasuda, *Corrosion*, 56, 887, 2000.
37. Moffitt, C. E., C. M. Reddy, Q. S. Yu, D. M. Wieliczka, R. Johnson, J. E. Deffeyes, and H. K. Yasuda, *Corrosion*, 56, 1032, 2000.
38. Yu, Q. S., C. M. Reddy, C. E. Moffitt, D. M. Wieliczka, R. Johnson, J. E. Deffeyes, and H. K. Yasuda, *Corrosion*, 57(9), 802, 2001.
39. Matsuzawa, Y. and L. Winterton, presentation at International Round Table on Plasma Interface Engineering, Surface Science and Plasma Technology, University of Missouri, Columbia, MO, 2002.
40. Dame, G. et. al., data to be published, 2010.

Index

A

Ablation
  competitive, 112
  description, 115
  gaseous species, 166
  metal from cathode surface, 94
  plasma sensitivity series, 119
  primary electron avalanche induced,
    94
  reactive species, 114
  shift, 116
AC, *see* Alternating current
AF, *see* Audio frequency
Alloying, 237
Alternating current (AC)
  DC cathodic polymerization and,
    154
  glow discharge, 155
  resistivity of LDPE film, 231
Antithrombogenic surface, 197
Audio frequency (AF), 7
  discharge, 7
  dissociation glow, 138
  MLCVD operation, 24, 34
  polymerization comparison, 82
  system pressure in, 182
Aurora borealis, 2, 96
Aurora CVD, 98

B

Bias potential, 8
Biocompatibility, *see also* Magneto-
    luminous chemical vapor
    deposition, applications of
  artificial material, 13
  biomedical electronics devices, 220
  carbon nanofilm, 228
  man-made material, 197, 201
  minimum perturbation theory of, 13
  MLCVD coatings, 197, 200
  polymers, 190

surface preparation for, 216
  sustainability, 13
Biomaterials
  bulk properties of materials, 198
  coagulation problems, 197
  ideal preparation tool, 196
  LCVD coating, 197
Biomimicking principle, 200

C

CAP principle, *see* Competitive ablation
    and polymerization principle
Cathode(s)
  bombardment of accelerated ions on,
    44–45
  breakdown voltage and, 49
  cathodic polymerization, 155
  deposition rate, 161
  dissociation glow, 8, 105
  E-coat, 238
  electric field, 61
  electron impact reactions, 7
  fall dark spaces, 34
  fall region, 7
  free electrons emission, 45
  glow, 30
  investigation of glow near, 36
  ions accelerated toward, 30, 53
  magnetron plasma sputter
    deposition, 11
  main electron flux, 92
  nonmagnetron, 75
  plasma polymer, 162, 164
  primary electron avalanche, 94
  radiant matter, 27
  surface state, 50
  toroidal glow surface, 174, 176
Chemical vapor deposition (CVD), 10
  aurora, 98
  description, 10
  reaction mechanism, 29
  thermal dissociation, 29

Competitive ablation and
    polymerization (CAP)
    principle, 112
  glow discharge treatment, 118
  influence, 116
  oxygen plasma treatment, 119
  schematic diagram, 113
Concepts, *see* Terms used and concepts
Corrosion
  gas–gas, 134
  protection
    aluminum alloy, 22, 118
    catastrophic damage to, 118
    galvanic, 23
    mainstream approach, 227
    oxides, 237
    principle applicable in, 2
    system approach interface
        engineering for, 21–24
  substrate metal, 231
CVD, *see* Chemical vapor deposition

D

Dark current, 51
DC, *see* Direct current
Deposition gas, 8
  dissociable, 85
  dissociation glow, 78, 154
  energy transfer, 105
  film growth rate and, 179
  gas phase breakdown, 29, 87
  hydrogen abstraction, 109
  oxygen, 133
  plasma phase of, 30
  RSGP mechanism, 107
  TMS, 106
  wall contamination and, 178
Deposition kinetics, *see* Operation
    parameters and deposition
    kinetics
DG, *see* Dissociation glow
Dielectric breakdown (gas phase),
    33–70
  argon glow, 37
  breakdown voltage, 49, 60, 65
  Coulombic attractive force, 60
  dark current, 51
  dark gas phase, 55

electric field, 53
electron accelerating dark space, 34
electron avalanche, 52, 60, 61
electronegativity of atom and
    efficiency of electron-impact
    reactions, 62–63
electron-impact reaction, 59, 62
experimental examination, 51–60
  breakdown process investigated,
      59–60
  breakdown voltage according to
      dark gas phase parameters,
      55–59
  breakdown voltage according
      to Townsend–Paschen
      hypothesis, 52–59
  factors not considered in Paschen–
      Townsend interpretation of gas
      phase breakdown, 55
  flaws in Paschen–Townsend
      treatment of gas phase
      breakdown, 54–55
  parameters influencing gas phase
      breakdown, 51–52
  Townsend theory of dielectric gas
      phase breakdown, 52–54
factors controlling transformation of
    gas phase, 61–62
  electron avalanche, 61
  parameters of reaction kinetics,
      61–62
  Townsend equation, 61
free electrons, 45, 50
gas phase breakdown as functions of
    system parameters, 63–69
gas system definition, 55
iN/Out Rule, 63
interfacial electron transfer, 46–51
  free electrons in surface state of
      cathode metal and breakdown
      current, 48–51
  static charge creation in contact
      electrification, 46–48
irradiation of cosmic ray, 45
luminous chemical vapor deposition,
    33–34
methane, plasma polymer of, 48
MLCVD, advantages of, 34
mono-atomic gas, 68

nondeposition gas and deposition gas, 35–43
  dissociation glow and ionization glow, 35–41
  interrelationship of dissociation glow and ionization glow, 42–43
parallel plate electrodes, breakdown studies, 54
Paschen curves, 56
Paschen–Townsend approach, 61
phase diagram, dielectric breakdown, 67
photo-emitting neutral species, 59
plasma polymerization reactors, breakdown process in, 57
plasma sheath, 55
reactor gas phase, 57
secondary electrons, 53, 60
source of electrons for electron avalanche, 44–46
  free electrons emission (primary electron emission principle by Yasuda), 45–46
  secondary electrons emission (Townsend's gas phase ionization hypothesis), 44–45
surface-state electrons, 51
system pressure, breakdown current and, 66
Townsend equation, 61
transition point pressure, 64
tribo-electric electron transfer, 52
trimethylsilane, 35, 38
Yasuda Parameter, 65, 68
Direct current (DC), 7
  argon, 33
  cathodic polymerization, 105, 154
  magnetic field, 71
  plasma polymerization, 133, 178
  plasma sputter coating, 91
  power source, 7
Dissociation glow (DG), 78, 162
  cathode surface, 82, 98
  description, 8
  discovery of, 29, 35
  dynamic steady state, 40
  energy level, 39
  hydrogen abstraction, 137

location, 138, 144
molecular, 155, 158
monomer gas in, 157
photo-emitting species in, 41
power source, 42, 174
RF discharge, 154
TMS, 37, 39, 79, 80
toroidal, 91

E

E-coat, 238
Effluent
  cleaning process, 1
  containment, 20
  environment-damaging, 1
  -free process, 18
  process yielding minimum, 15
  release, 1
EIS, *see* Electrochemical impedance spectroscopy
Electrochemical impedance spectroscopy (EIS), 221
  Bode plots, 224, 225, 226
  freestanding Parylene C film and, 223
  interfacial factors, 222
  principle, 221
Electron(s)
  accelerating dark space, 34
  affinity, 12
  attachment to dissociation gas, 79
  avalanche, 52, 60, 61
  binding energy, 150
  emission of from cathode surface, 3
  free, 45, 50
  -impact reaction, 2, 35, 59, 62
  negativity, 133
  secondary, 53, 60
  surface-state, 51
  temperature, 71
  transfer, tribo-electric, 52
  unpaired, 108
  valence-level, 45
Electronegativity, 8
Electroneutrality, 6
Environmental remediation process, cost of, 1, 15

F

Faraday Cage, 220
Free radicals(s), 132
    deposited plasma polymer, 132
    polymerization, 101
        acrylic acid, 172
        functional groups, 102
        luminous gas phase, 101
        monomers, 133
        Parylene, 223
        steps coupled in, 146
    recombination, 104, 107
    trapped, 174
Front-end approach, 1, 15, *see also* Green
        deposition coating (nanofilms)
Front-end super green process, 3

G

Gas
    deposition, 8
        dissociable, 85
        dissociation glow, 78, 154
        energy transfer, 105
        film growth rate and, 179
        gas phase breakdown, 29, 87
        hydrogen abstraction, 109
        oxygen, 133
        plasma phase of, 30
        RSGP mechanism, 107
        TMS, 106
        wall contamination and, 178
    electric discharge of, 7–8
        audio frequency discharge, 7
        cathode fall, 7
        deposition gas, 8
        direct current discharge, 7
        dissociation glow, 8
        electronegativity, 8
        ionization glow, 7
        nondeposition gas, 8
        radio frequency discharge, 7
        self-bias and bias potential, 8
    equation, 8
    ionization, 9, 44
    mono-atomic, 68
    nondeposition, 3, 8
        description, 8

    dielectric breakdown of, 33, 35
    flow system pressure, 9
    mono-atomic, 52, 68
    plasma polymerization, 29
    system pressure, 134
Gas phase, 8–10, *see also* Dielectric
        breakdown
    breakdown, 45, 49, 52, 183
    closed-system, 109
    dark, 55
    equation, 54, 55
    factors controlling transformation of,
        61–62
    flow rate, volume flow rate, mass
        flow rate, 9
    gas equation, 8–9
    ionization hypothesis, 44
    luminous, see Luminous gas phase
    magnetic field, 9–10
    paramagnetic, 89
    photo emitting excited neutrals, 9
    reactor, 57
Glow film, 30
Green deposition coating (nanofilms),
        15–25
    acidic polymer, 17
    atmospheric electrical discharge, 21
    contact lens, 16
    corrosion protection, 22
    front-end approach and rear-end
        approach, 15–21
        cost of coating processes, 18–21
        layer-by-layer coating, 16–18
        low-pressure magneto-luminous
            chemical vapor deposition
            coating, 16
    glow discharge polymerization, 24
    green processing, 23
    ion vapor deposition, 22
    MLCVD coating, 15, 18
    plasma SAIE, 21
    polymer–polymer salt, 18
    radio frequency plasma processes, 24
    removal of excess solution, 18
    system approach interface
        engineering, 21–24
        corrosion protection, 21–24
        magneto-luminous CVD, 24

wastewater treatment cost, 19, 20
wet chemistry process, 19
zinc, 23
Green process(es), *see also* Green
    deposition coating
    aim of, 1
    approaches, 1
    cost, 1
    front-end, 3, 171, 237
    rear-end, 3
    super-, 238
    ultimate, 18

H

HDPE, *see* High-density polyethylene
HEF, *see* Hexafluoroethane
Hexafluoroethane (HFE), 119, 120
High-density polyethylene (HDPE), 231
Hydrocarbon(s)
    deposition characteristics, 127
    LCVD, 143
    monomers, 109, 124
    saturated, 104, 106, 137
Hydrophilic surface, 187, 193, 216

I

IG, *see* Ionization glow
iN/Out Rule
    broken-down gas phase, 90
    description, 63
    N atom, 115
    principle, 63
    surface oxidation, 76
Ionization
    argon, 77
    cause, 30
    consequence, 35
    dissociation product, 106
    electron-impact, 38, 62
    energy, 28, 37, 113
    glow (IG), 78, 162
        argon, 37, 73
        description, 7
        dynamic steady state, 40
        energy level, 39
        negative glow region, 77

photo-emitting species in, 41
power source and, 42
RF discharge, 154
TMS, 80
mono-atomic gas, 68
Ion vapor deposition (IVD), 22, 23
IVD, *see* Ion vapor deposition

K

Kinetic path length, 144, 148, 177

L

Layer-by-layer (LBL) process, 16–18
    coating line, 17
    contact lens surface, 214
    cost, 18, 19
    description, 16
    wastewater treatment, 20
LBL process, *see* Layer-by-layer process
LCVD, *see* Luminous chemical vapor
    deposition
LDPE, *see* Low-density polyethylene
Low-density polyethylene (LDPE), 231,
    236
Luminous chemical vapor deposition
    (LCVD), 10, *see also* Magneto-
    luminous chemical vapor
    deposition
    biomaterials, 197
    cathodic, 230
    closed-system, 23, 165, 168
    coating, costs, 19
    description, 10
    development hurdles, 22
    flow system, system pressure, 165
    hydrocarbon, 143
    loading factor, 137
    low-pressure operation of, 178
    material formation in, 137, 155, 162
    nanofilm barrier characteristics, 204
    reaction characteristics, 104
    reactive species, 145
    reactor cost, 21
    system dependent nature of, 138
Luminous gas, *see* Polymer formation
    mechanism

Luminous gas phase, 28–30, *see also*
    Magnetic field, influence of on
    luminous gas phase
  argon, 28
  cathode glow, 30
  deposition on floating substrate in,
    159–162
  deposition kinetics, 138–142
  dielectric breakdown of gas phase, 28
  dissociation glow, 29
  glow film, 30
  ionized gas, 28
  loss of gas species, 29
  luminous chemical vapor deposition,
    29
  methane, 28
  moving substrates in, 174
  neutral species, 30
  plasma polymerization, gases used
    in, 29
  polymer formation mechanism, 105
  radiant matters, 28
  repeating step growth
    polymerization, 29

M

Magnetic field, influence of on luminous
    gas phase, 71–99
  aurora CVD, 98
  aurora inception mechanism, 98
  breakdown current, 84
  breakdown voltages, 82, 83
  cathodic dark place, 71
  edge effect, 76
  electrons in electric field and in
    magnetic field, 88–90
  electron temperature, 71
  fluorine-containing gases, 80
  gas discharge without magnetic
    field, 88
  gas molecules, energy level of, 90
  gas phase, paramagnetic, 89
  influence of magnetic field on
    dielectric breakdown of gas
    phase, 81–88
  ionized gas phase, 78
  Langmuir probe measurement, 72
  luminous gas phase, 78

  magnetic field initiation of luminous
    gas phase, 94–98
  collisions of gas molecules with
    electrons in magnetic field,
    94–96
  potential mechanism for
    inception of aurora borealis,
    96–98
  magnetron discharge, 74
  magnetron gas phase breakdown,
    implications of, 90–94
  magnetron chemical vapor
    deposition versus magnetron
    sputtering of cathode metal,
    90–92
  magnetron discharge sputtering,
    92–94
  magnetron plasma polymerization,
    94
  momentum exchange mechanism,
    92
  mono-atomic gas, 85
  negative glow, 77
  nonequlibrium plasma, 72
  oxygen-sensing devices, 89
  paramagnetic oxygen, 89
  plasma state, 71
  primary electron avalanche, 94
  roles of electrons, 71–76
  changes in distribution profile
    of Te and Ne in luminous gas
    phase, 71–72
  magnetron cathode and
    nonmagnetron anode, 73–74
  nonmagnetron cathode and
    magnetron anode, 75–76
  shaping of negative glow near
    magnetron anode, 76–81
  influence of magnetron anode
    on glow characteristics of
    deposition gas, 78–80
  negative glow of argon, 76–78
  reexcitation of photo-emitting
    species by low-energy
    electrons, 77–78
  shift of dissociation glow from
    cathode surface to gas phase,
    81
  THEMIS Project (NASA), 96

toroidal glow, 93
transition point, 82
Magneto-luminous chemical vapor
        deposition (MLCVD), 1, 15,
        171–185
    advantages, 88, 134, 185
    Ar glow discharge, 172
    audio frequency discharge, 182
    audio frequency power sources, 24
    aurora inception mechanism, 98
    bell jar–type reactor, 65
    broken-down gas phase, 183
    coating system cost, 18
    confined luminous gas phase in low
        pressure, 177–178
    contact lens coating line, 17
    deposition rate, 181
    discharge current, 184
    domain, 86, 171–175
    equipment cost, 173
    film growth rate, 180
    glow volume, confining of, 177
    imperturbable surface state, 171
    industrial-scale plasma reaction
        system, 172
    low-pressure, coating, 16
    luminous gas phase, moving
        substrates in, 174
    magnetic field, 181
    magnetron discharge sputtering, 92
    nanofilm, 2, 200
    operation modes, 175
    oxidative etching, 176
    oxygen in, 133, 176
    polymer formation and deposition in
        low pressure, 178–185
        deposition kinetics in domain M,
        178–182
        pressure dependence of
        deposition rate, 182–184
        small grain size and uniform
        smooth surface, 184–185
        strong adhesion to substrate
        surface, 185
    power input, 178
    prerequisites in comprehending, 34
    radio frequency discharge, 179
    RSGP mechanism, 174
    significance of, 2
simple organic molecules in, 103
substrate, distribution patterns, 175
toroidal glow surface without
        deposition, 175–177
Magneto-luminous chemical vapor
        deposition (MLCVD),
        applications of, 187–240
    agar-agar gel, 194
    alloying, 237
    amphoteric surface, 187
    antithrombogenic surface, 197
    biocompatibility, 190, 197
    biomaterials, 197
    biomimicking approach, 200
    blood compatibility, 197
    chromate conversion coating, 238
    coagulation problems, 197
    Columbic attractive force, 231
    corrosion protection, 238
    diffusion transport characteristics,
        231
    E-coat, 238
    electrochemical corrosion inhibitors,
        227
    electrochemical impedance
        spectroscopy, 221
    Faraday Cage, 220
    FDA regulation, 213
    gel–air interface, 192
    green process, 237
    heparinized surface, 197
    hydrophilic surface, 187
    hydrophobic recovery, 191, 201
    imperturbable aspect of surface, 197
    implantation of imperturbable
        surface state on substrate,
        187–200
        factors involved in interface,
        189–190
        implantation of surface state
        of MLCVD nanofilm onto
        material surface, 196–200
        molecular configuration versus
        surface configuration, 190–196
        surface dynamic change, 187–189
    interface engineering for adhesion of
        coating, 230–238
        salt diffusion versus salt
        intrusion, 230–236

system approach interface engineering, 236–238
layer-by-layer process, 214
migration, 189
minimum perturbation principle, 200
MLCVD nanofilm for
 biocompatibility, 200–230
 encapsulation of metallic stent, 219–220
 encapsulation of silicone contact lens, 201–211
 encapsulation of silicone/ hydrogel contact lens, 211–219
 imperturbable surface state and biocompatibility, 200–201
 plasma system approach interface engineering for biomedical electronics devices, 220–228
 unique features of MLCVD amorphous carbon nanofilm for biocompatibility, 228–230
nanofilm, 200
Parylene C, 224
polymer coated stent, 220
poly(oxy ethylene), 198
reference state, 187
salt intrusion, 231
sample holder, 218
substrate metal, corrosion of, 231
surface configuration, 189, 191
surface dynamics, 189, 191, 192, 201
surface reconstruction, 191, 201
water droplet shape, 193
water-immersed samples, 188
water-insensitive adhesion, 227
x-ray photoelectron spectroscopy, 187
zinc-phosphated steel, 238
Magnetron plasma sputter deposition, 11
Mass flow controller, 135
Migration, 121, 189
Minimum perturbation principle, 200
Minimum perturbation theory of biocompatibility, 13
MLCVD, *see* Magneto-luminous chemical vapor deposition
Monomer(s)
 characteristic, 138
 -deficient domain, 143, 153

dissociation glow, 157
flow rate, 134, 135–136
free-radical polymerization, 101, 133
gas
 dissociation glow, 155, 158
 extinction, 182
 steady-state flow of, 131
hydrocarbon, 109, 124
hydrophilic, 211
iN/Out Rule, 63
molecular configuration of, 2
molecular weight, 139
molecules, dissociation of, 122–124
plasma polymerization, 146
-polymer conversion ratio, 136
saturated, 102
selection, 174
TMS, 166
types, classification, 104
vapor, 29
vinyl, 101, 103

N

Nanofilms, *see also* Green deposition coating
 barrier characteristics, 204
 carbon, 228
 MLCVD, 200
NASA, *see* National Aeronautics and Space Administration
National Aeronautics and Space Administration (NASA), 2
 experiments, 97
 THEMIS Project, 96
Nondeposition gas, 3
 description, 8
 dielectric breakdown of, 33, 35
 flow system pressure, 9
 mono-atomic, 52, 68

O

Operation parameters and deposition kinetics, 129–169
 cathodic polymerization, 155, 161
 deposition-growth polymerization, 151
 deposition kinetics, 136–144

mass balance in flow deposition
system, 136–138
normalized deposition rate,
142–144
normalized energy input
parameter to luminous gas
phase, 138–142
deposition rate dependence, 155
dielectric substrate, 143
discharge wattage, 143
dissociation glow, 157
electrical discharge system, 139
film thickness growth rate, 156
flow system LCVD, system pressure,
165
free radicals, 132, 133
gas yield, 137
glow discharge, 132, 155
kinetic path length, 144
mass flow controller, 135
material formation in LCVD, 137
monomer characteristic, 138
monomer-deficient regime, 141, 143
monomer gas, steady-state flow of,
131
operation time, 129
partition of deposition on electrode
and deposition on surface in
gas phase, 154–168
cathodic polymerization versus
polymerization in negative
glow of DC discharge, 154
DC plasma polymerization in
closed system, 165–168
deposition on floating substrate in
luminous gas phase, 159–162
deposition profile on electrode,
154–159
role of anode in DC cathodic
polymerization, 162–165
perfluoropropene, RF plasma
polymerization of, 148
plasma deposition step, 146
plasma polymer coating in vacuum,
130
plasma polymerization process,
129–136
control of monomer flow rate,
135–136

flow rate and system pressure of
gas, 134–135
repeating step growth
polymerization mechanism,
129–133
plasma polymerization reactor, 145
plasma polymers and plasma
polymerization, 144–154
type A and type B plasma
polymers, 144–153
utilities of type A and type B
plasma polymers, 153–154
polymer yield, 137
postplasma graft polymerization, 133
pump-out rate, 136
reactor volume, 134
system dependent nature of LCVD,
138
TMS coating thickness, 167
TMS deposition rate, 158
tubular reactor, 148
Oxide
aluminum, 22
aluminum fluoride-, 120
chromium, 238
ethylene, 111
layer removal, 118
SAIE system, 237

P

PACVD, *see* Plasma assisted CVD
Paschen–Townsend gas phase
ionization theory, 54
PCVD, *see* Plasma chemical vapor
deposition
PECVD, *see* Plasma enhanced CVD
Perfluorbutyltetrhydrofuran (PFBTHF),
145, 146
PFBTHF, *see*
Perfluorbutyltetrhydrofuran
Plasma, 5–7
assisted CVD (PACVD), 10
chemical vapor deposition (PCVD),
10
deposition plasma and
nondeposition plasma, 6
deposition step, 146
enhanced CVD (PECVD), 10

equilibrium and nonequilibrium
        plasma, 5–6
low-pressure (low-temperature)
        plasma, 6
magneto-plasma, 7
phase, 27–28
    blood plasma, 27
    ionized gas phase, 27
    low-pressure plasma, 28
    radiant matter, 27
reactive plasma and nonreactive
        plasma, 6
SAIE (P-SAIE), 21
    advantages, 21–22, 23
    case, 22
    corrosion protection, 227, 238
    importance, 221
    objects using, 24
sheath
    dark gas phase and, 55
    description, 6
    energy input, 33
Plasma polymerization, 10
    direct current, 133, 178
    equation, 155
    gas, 29
    low-pressure, 33
    magnetron, 11, 94
    material formation, 10
    monomer, 146
    operation parameters and, 129–136
        control of monomer flow rate,
            135–136
        flow rate and system pressure of
            gas, 134–135
        repeating step growth
            polymerization mechanism,
            129–133
    reactors, breakdown process in, 57
    TMS, 123
POE, see Poly(oxy ethylene)
Polymer(s)
    biocompatibility, 190
    biological system and, 201
    bulk properties of, 2
    chain mobility, 190
    chemical moieties, 191
    coated stent, 220
    deposition from TMS, 123

-forming species, 155
hydrophilic, 196, 211
interfacial tension, 189
ionic, 17
LBL coating, 16
low-crystallinity, 221
magnetron plasma polymerization,
        94
–metal interface, 236
perturbability, 191
plasma
    categories, 138
    cathodic, 162, 164
    contact lens, 185, 208
    free radicals, 132
    hydrophobic, 227, 232
    luminous chemical vapor
        treatment, 168
    main surface constituent, 133
    methane, 48, 204, 206
    monomer flow rate and, 135
    N atoms, 63
    perfluoropropene, 149, 150
    reactive species, 146
    substrate and, 129
    temperature dependence, 145,
        148
    tetramethyldisiloxane, 140, 141
    toroidal glow surface, 175, 176
    type A, 153
    vacuum system, 130
    wall contamination, 82, 178
–polymer salt, 18
silicone, 202, 203, 211, 214
surface configuration, 199
surface state, 2
water-soluble, 192
yield, 137
Polymer formation mechanism
        (luminous gas), 101–128
    cathodic polymerization, 118
    closed-system gas phase, 109
    closed-system polymerization, 123
    competitive ablation and
        polymerization principle,
        112–116
    contaminated reactor, 120
    dependence of polymer formation on
        operation parameters, 124–127

deposition gas, 106
deposition plasma, species identified in, 111
di-free radicals, 107
dissociation of monomer molecules, 122–124
dissociation processes, 108
electron-impact dissociation, 102, 108
energy-deficient domain, 126
fabric porosity, 116
F-containing oligomers, 119, 120
free-radical polymerization and free-radical polymer formation, 101–102
gaseous reactive species, 114
hydrogen production, 109
influence of unaccounted factors, 116–122
    reactor wall contamination, 118–122
    substrate material, 116–118
initiator, concentration of, 101
ionization, 106
luminous gas phase, polymer formation mechanism, 105
mono-free radicals, 107
monomer-deficient domain, 126
perfluorocarbons, 125
plasma sensitivity, 112
pulsed radio frequency, 111
repeating step growth polymerization mechanism, 102–112
secondary ion mass spectroscopy, 115
styrene, plasma polymerization of, 105
system pressure, 124
tetrafluoroethylene, 125
tricyclic RSGP mechanism, 110
x-ray photoelectron spectroscopy, N detected by, 114–115
Polymerization, material formation and, 10–11
chemical vapor deposition, 10
deposition mechanism, 11
luminous chemical vapor deposition, 10

magneto-luminous CVD, magneto-plasma CVD, magnetron-plasma polymerization, 11
magnetron plasma sputter deposition, 11
material formation mechanism, 11
plasma chemical vapor deposition, 10
plasma enhanced CVD; plasma assisted CVD, 10
plasma polymerization, 10
Poly(oxy ethylene) (POE), 198
P-SAIE, *see* Plasma SAIE

R

Radiant matter, 27
Radio frequency (RF), 7
    discharge, 7
    glow discharge, 24, 109
    low-pressure plasma polymerization, 33
    plasma polymerization, 133
    RSGP mechanism and, 111
Rear-end approach, 1, 15, *see also* Green deposition coating
Reference state, 37, 187, 201
Repeating step growth polymerization (RSGP), 29, 102, 182
RF, *see* Radio frequency
RSGP, *see* Repeating step growth polymerization

S

SAIE, *see* System approach interface engineering
Salt
    diffusion, 230, 231
    intrusion, 223, 230, 231, 236
    ions, 231
    permeation, 231
    polymer–polymer, 18
Secondary electron emission, 53, 60, 63
Secondary ion mass spectroscopy (SIMS), 115
Self-bias potential, 8
Silicone contact lens, encapsulation of (MLCVD nanofilm), 201–211
    dye penetration test, 203–205
    effect of coating thickness, 205–207

effect of power input level, 207–209
overall effects of MLCVD coating,
  209–211
Silicone/hydrogel contact lens,
  encapsulation of (MLCVD
  nanofilm), 211–219
  advantages and disadvantages of
    silicone/hydrogel contact
    lenses, 211–216
  industrial-scale batch and
    continuous operation of
    MLCVD CH4 coating, 216–219
SIMS, *see* Secondary ion mass
  spectroscopy
Surface
  configuration, 191
    change of, 189, 190, 197
    contacting medium and, 201
    description, 12
    imperturbability of, 13
    POE surface, 199
    surface characteristics and, 2
  dynamics, 187
    description, 187, 189
    terms, 191, 201
  interface and, 11–13
    imperturbable surface state, 12–13
    interface, 11
    interfacial electron transfer, 12
    molecular configuration, 12
    surface, 11
    surface configuration, 12
    surface state, 12
    surface static charge, 12
  reconstruction, 187
  state, 12
    biocompatibility and, 13
    boundary, 237
    breakdown of, 232, 234
    cathode, 49
    change of, 2, 203
    electrons in, 12, 51
    free electrons emission from, 45
    imperturbable, 2, 171, 187, 200
    low-pressure gas phase, 6
    semicrystalline polymer, 230
    silicone lens, 209
    substrate material, 196

System approach interface engineering
  (SAIE), 21, 236–238
  conventional, 237–238
  corrosion protection, 21
  MLCVD-, 238
  need for, 236
  plasma, 21, 221, 237–238
  rear-end approach, 23

T

TEF, *see* Threshold electric field
Terms used and concepts, 5–13
  audio frequency discharge, 7
  biocompatibility, 13
    artificial material, 13
    minimum perturbation theory, 13
    sustainability, 13
  chemical reaction, parameter
    defining, 9
  electric discharge of gas, 7–8
    audio frequency discharge, 7
    cathode fall, 7
    deposition gas, 8
    direct current discharge, 7
    dissociation glow, 8
    electronegativity, 8
    ionization glow, 7
    nondeposition gas, 8
    radio frequency discharge, 7
    self-bias and bias potential, 8
  electroneutrality, 6
  flow system, pressure in, 9
  gas law, 9
  gas phase, 8–10
    flow rate, volume flow rate, mass
      flow rate, 9
    gas equation, 8–9
    ionized gas, 9
    luminous gas, 9
    magnetic field in gas phase, 9–10
    photo emitting excited neutrals, 9
  low-pressure plasma, 6
  plasma, 5–7
    deposition plasma and
      nondeposition plasma, 6
    equilibrium and nonequilibrium
      plasma, 5–6

low-pressure (low-temperature)
plasma, 6
magneto-plasma, 7
plasma sheath, 6
reactive plasma and nonreactive
plasma, 6
polymerization and material
formation, 10–11
chemical vapor deposition, 10
deposition mechanism, 11
luminous chemical vapor
deposition, 10
magneto-luminous CVD,
magneto-plasma CVD,
magnetron-plasma
polymerization, 11
magnetron plasma sputter
deposition, 11
material formation mechanism, 11
plasma chemical vapor
deposition, 10
plasma enhanced CVD; plasma
assisted CVD, 10
plasma polymerization, 10
radio frequency discharge, 7
surface and interface, 11–13
imperturbable surface state, 12–13
interface, 11
interfacial electron transfer, 12
molecular configuration, 12
surface, 11
surface configuration, 12
surface state, 12
surface static charge, 12
THEMIS Project (NASA), 96
Threshold electric field (TEF), 52
dark gas phase, 55
dielectric breakdown, 55
electron avalanche and, 52
plot, 56, 57
TMS, *see* Trimethylsilane
Townsend equation, 61
Townsend–Paschen hypothesis, 52–59
Trimethylsilane (TMS), 119
DC glow discharge, 166
deposition gas, 106, 158
dielectric breakdown, 2
dissociation glow, 29, 35, 78
LCVD operation cost, 23

MLCVD use, 103
plasma deposition of, 119
plasma polymerization of, 123

V

Vacuum
deposition processes, 2
equipment, cost, 20
plasma polymer coating in vacuum,
130
processing, 2, 20
important aspects, 20
nonfamiliarity of, 22
pump oil change, 120
surface characteristics measured
with, 201
system cost, 172
water removal, 215
Vinyl monomers, 101, 103

W

Wastewater treatment cost, 19, 20

X

XPS, *see* X-ray photoelectron spectroscopy
X-ray photoelectron spectroscopy (XPS),
133
Al alloy surfaces, 120
angular dependence, 188, 189
atomic ratio, 124, 187
C/Si ratios, 168
F counts, 117
fluorine-containing contaminants, 119
N detected by, 114–115
O atom found by, 133, 187
perfluoropropene, 148
polymer deposition from TMS, 123

Y

Yasuda parameter, 65, 68

Z

Zinc
alloy, 23
-phosphated steel, 238